Contents

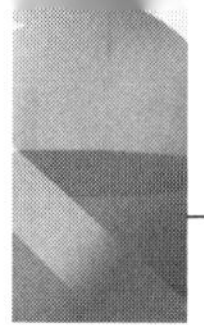

Introduction

About this guide

This unit guide is the first in a series of two, which together cover the whole OCR AS chemistry specification. This guide is written to help you prepare for the **Unit F321: Atoms, Bonds and Groups** examination.

This **Introduction** provides advice on how to use the guide, together with suggestions for effective revision.

The **Content Guidance** section gives a point-by-point description of all the facts you need to know and concepts that you need to understand for Unit F321. It aims to provide you with a basis for your revision. However, you must also be willing to use other sources in your preparation for the examination.

The **Question and Answer** section shows you the sort of questions you can expect in the unit test. It would be impossible to give examples of every kind of question in one book, but the questions used should give you a flavour of what to expect. Each question has been attempted by two candidates, candidate A and candidate B. Their answers, along with the examiner's comments, should help you see what you need to do to score a good mark — and how you can easily *not* score marks, even though you probably understand the chemistry.

What can I assume about the guide?

You can assume that:

- the topics covered in the Content Guidance section relate directly to those in the specification
- the basic facts you need to know are stated clearly
- the major concepts you need to understand are explained
- the questions at the end of the guide are similar in style to those that will appear in the unit test
- the answers supplied are genuine, combining responses commonly written by candidates
- the standard of the marking is broadly equivalent to the standard that will be applied to your answers

What can I *not* assume about the guide?

You must *not* assume that:

- every last detail has been covered
- the way in which the concepts are explained is the *only* way in which they can be presented in an examination (often concepts are presented in an unfamiliar situation)

- the range of question types presented is exhaustive (examiners are always thinking of new ways to test a topic)

So how should I use this guide?

The guide lends itself to a number of uses throughout your course — it is not *just* a revision aid.

The Content Guidance is laid out in sections that correspond to those of the specification for Unit F321, so that you can use it:
- to check that your notes cover the material required by the specification
- to identify strengths and weaknesses
- as a reference for homework and internal tests
- during your revision to prepare 'bite-sized' chunks of material rather than being faced with a file full of notes

The Question and Answer section can be used to:
- identify the terms used by examiners in questions, and what they expect of you
- familiarise yourself with the style of questions you can expect
- identify the ways in which marks are lost, as well as the ways in which they are gained

Study skills and revision techniques

All students need to develop good study skills. This section provides advice and guidance on how to study AS chemistry.

Organising your notes

Chemistry students often accumulate a large quantity of notes, so it is useful to keep these in a well-ordered and logical manner. It is necessary to review your notes regularly, maybe rewriting the notes taken during lessons so that they are clear and concise, with key points highlighted. You should check your notes using textbooks, and fill in any gaps. Make sure that you go back and ask your teacher if you are unsure about anything, especially if you find conflicting information in your class notes and textbook.

It is a good idea to file your notes in specification order using a consistent series of headings. The Content Guidance section can help you with this.

Organising your time

When organising your time, make sure that you plan carefully, allowing enough time to cover all of the work. It sounds easy, but it is one of the most difficult things to do. There is considerable evidence to show that revising for 2–3 hours at a time is counterproductive and that it is much better to work in short, sharp bursts of between 30 minutes and an hour.

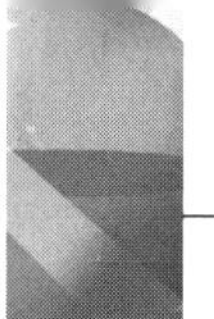

Preparation for examinations is personal. Different people prepare, equally successfully, in different ways. The key is being honest about what *works for you.*

Whatever your style, you must have a plan. Sitting down the night before the examination with your notes and a textbook does not constitute a revision plan — it is just desperation — and you must not expect a great deal from it. Whatever your personal style, there are some things you *must* do and some you *could* do.

The scheme outlined below is a suggestion as to how you might revise Unit 1 over a 3-week period. The work pattern shown is fairly simple. It involves revising and/or rewriting a topic, and then over the next few days going through it repeatedly but never spending more than 30 minutes at a time. When you are confident that you have covered all areas, start trying to answer questions from past papers or this guide's Question and Answer section. Mark them yourself and seek help with anything that you are not sure about.

Day	Week 1	Week 2	Week 3
	Each revision session should last approximately 30 minutes		
Mon	**30 mins** on atoms, moles and equations	Reread all your summary notes at least twice	You have now revised all of Unit 1 and have attempted questions relating to each topic. Make a list of your weaknesses and ask your teacher for help. Reread all your summary notes at least twice. Ask someone to test you
Tue	**20 mins** on acids and redox **10 mins** on atoms, moles and equations	Using past papers or other question sources, try a structured question on Module 1: **Atoms and reactions** Mark it and list anything you do not understand Allow about 30 minutes	Using past papers or other question sources, try a relevant question that requires extended writing (essay-type questions) from any of the topics Mark it and list anything you do not understand Allow about 30 minutes
Wed	**15 mins** on electron structure **10 mins** on acids and redox **5 mins** on atoms, moles and equations	Using past papers or other question sources, try a structured question on Module 2: **Electrons, bonding and structure** Mark it and list anything you do not understand Allow about 30 minutes	Reread all your summary notes at least twice Concentrate on the weaknesses that you identified on Monday (by now you should have talked to your teacher about them) Ask someone to test you

Day	Week 1	Week 2	Week 3
Thu	**15 mins** on bonding and structure **10 mins** on electron structure **5 mins** on acids and redox; atoms, moles and equations	Using past papers or other question sources, try a structured question on Module 3: **The periodic table** Mark it and list anything you do not understand Allow about 30 minutes	Collect together about five structured questions and one extended-answer question covering all six topics, and try them under exam conditions Allow 60 minutes Mark them and list anything you do not understand
Fri	**10 mins** on periodicity **10 mins** on bonding and structure **5 mins** on electron structure; acids and redox; atoms, moles and equations	Using past papers or other question sources, try two structured questions on Module 1: **Atoms and reactions** Mark them and list anything you do not understand Allow about 30 minutes	Reread all your summary notes at least twice Concentrate on the weaknesses that you identified on Monday (by now you should have talked to your teacher about them) Ask someone to test you
Sat	**15 mins** on group 2 and group 7 **10 mins** on periodicity **5 mins** on bonding and structure **5 mins** on electron structure; acids and redox; atoms, moles and equations	Using past papers or other question sources, try two structured questions on Module 2: **Electrons, bonding and structure** Mark them and list anything you do not understand Allow about 30 minutes	Try a complete past exam-paper under exam conditions. Past papers are available on OCR's website at **www.ocr.org.uk.** Check your answers against the mark schemes
Sun	**10 mins** on group 2 and group 7 **20 mins** on periodicity; bonding and structure; electron structure; acids and redox; atoms, moles and equations	Using past papers or other question sources, try two structured questions on Module 3: **The periodic table** Mark them and list anything you do not understand Allow about 30 minutes	Try a complete past exam-paper under exam conditions. Past papers are available on OCR's website at **www.ocr.org.uk**. Check your answers against the mark schemes

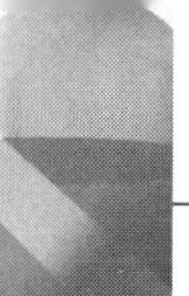

This revision timetable may not suit you, in which case write one to meet your needs. It is only there to give you an idea of how one might work. The most important thing is that the grid at least enables you to see what you should be doing and when you should be doing it. Do not try to be too ambitious — *little and often is the best way.*

It would of course be sensible to put together a longer rolling programme to cover all your AS subjects. Do *not* leave it too late. Start sooner rather than later.

Things you *must* do

- Leave yourself enough time to cover *all* the material.
- Make sure that you actually *have* all the material to hand (use this guide as a basis).
- Identify weaknesses early in your preparation, so that you have time to do something about them.
- Familiarise yourself with the terminology used in examination questions.

Things you *could* do to help you learn

- Copy selected portions of your notes.
- Write a precis of your notes, which includes all the key points.
- Write key points on postcards (carry them round with you for a quick revise during a coffee break).
- Discuss a topic with a friend who is studying the same course.
- Try to explain a topic to someone *not* on the course.
- Practise examination questions on the topic.

Approaching the unit test

Terms used in examination questions

You will be asked precise questions in the examinations, so you can save a lot of valuable time (as well as ensuring you score as many marks as possible) by knowing what is expected. Terms most commonly used are explained below.

Define

This is intended literally. Only a formal statement or equivalent paraphrase is required.

Explain

This normally implies that a definition should be given, together with some relevant comment on the significance or context of the term(s) concerned, especially where two or more terms are included in the question. The amount of supplementary comment intended should be determined by the mark allocation.

State

This implies a concise answer with little or no supporting argument.

Describe
This requires you to state in words (using diagrams where appropriate) the main points of the topic. It is often used with reference either to particular phenomena or to particular experiments. In the former instance, the term usually implies that the answer should include reference to (visual) observations associated with the phenomena. The amount of description intended should be interpreted according to the indicated mark value.

Deduce or predict
This implies that you are not expected to produce the required answer by recall but by making a logical connection between other pieces of information. Such information may be wholly given in the question or may depend on answers given in an earlier part of the question. 'Predict' also implies a concise answer, with no supporting statement required.

Outline
This implies brevity, i.e. restricting the answer to essential detail only.

Suggest
This is used in two main contexts. It implies either that there is no unique answer or that you are expected to apply your general knowledge to a 'novel' situation that may not be formally in the specification.

Calculate
This is used when a numerical answer is required. In general, working should be shown.

Sketch
When this is applied to diagrams, it implies that a simple, freehand drawing is acceptable. Nevertheless, care should be taken over proportions, and important details should be labelled clearly.

Content Guidance

Unit F321, Atoms, Bonds and Groups, includes basic information that you will need for the whole of the AS/A2 course. A good understanding of this unit, particularly the mole concept and the use of balanced equations, provides you with a solid foundation on which to build.

This Content Guidance section is a student's guide to Unit F321, which is divided into three parts. In **Atoms and reactions**, the main topics are:

- atomic structure
- moles and equations
- the mole and the Avogadro constant
- acids
- redox

In **Electrons, bonding and structure**, the main topics are:

- electron structure
- bonding and structure

In **The Periodic Table**, the main topics are:

- periodicity
- group 2
- group 7

This Content Guidance section includes all the relevant key facts required by the specification, and explains the essential concepts.

Atoms and reactions

Atomic structure

Protons, neutrons and electrons

Atoms consist of a nucleus containing protons and neutrons, which is surrounded by electrons.

Particle	Relative mass	Relative charge	Distribution
Proton, p	1	+1	Nucleus
Neutron, n	1	0	Nucleus
Electron, e	1/1836	−1	Orbits/shells

The atomic and the mass numbers (see below) can be used to deduce the number of protons, neutrons and electrons in atoms and ions. Atoms are neutral and contain the same number of protons as electrons. Positive ions have lost electrons and hence have more protons than electrons, while negative ions have gained electrons, so they have fewer protons than electrons.

- The atomic number tells us the number of protons in an atom.
- The mass number tells us the number of protons plus the number of neutrons in an atom.
- The mass number minus the atomic number tells us the number of neutrons in an atom.

$^{31}_{15}P$ — 15 p, 15 e, 16 n

$^{32}_{15}P$ — 15 p, 15 e, 17 n

$^{24}_{12}Mg^{2+}$ — 12 p, 10 e, 12 n

$^{32}_{16}S^{2-}$ — 16 p, 18 e, 16 n

^{31}P and ^{32}P are isotopes

As you can see from the diagram above, ^{31}P and ^{32}P are isotopes of phosphorus. Isotopes are atoms of the same element with different masses — they have the same number of protons (and electrons) but different numbers of neutrons. Chlorine has two isotopes: ^{35}Cl and ^{37}Cl.

Relative masses

Most Atoms, Bonds and Groups papers ask for at least one or two definitions. Remember that 'Define' is intended literally. Only a formal statement or equivalent paraphrase is required.

Definitions of relative isotopic mass and relative atomic mass are often required, and each one is based on the carbon-12 scale.

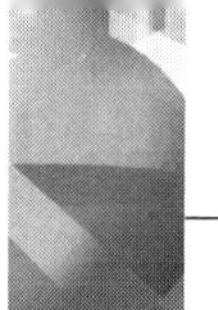

A similar definition is needed for each:

- The **relative isotopic mass** is defined as the mass of an atom/isotope of the element compared with one-twelfth of the mass of a carbon-12 atom.
- The **relative atomic mass** of an element is defined as the weighted mean mass of an atom of the element compared with one-twelfth of the mass of a carbon-12 atom.

It is possible to use this basic definition and amend it slightly to create definitions of the relative molecular mass and the relative formula mass:

- The **relative molecular mass** is defined as the weighted mean mass of a molecule compared with one-twelfth of the mass of a carbon-12 atom.
- The **relative formula mass** is defined as the weighted mean mass of a formula unit compared with one-twelfth of the mass of a carbon-12 atom.

You will be expected to calculate the relative atomic mass from data, where you will be given information about each isotope and told the relative abundance of each isotope.

Example 1

A sample of chlorine contains about 75% ^{35}Cl and 25% ^{37}Cl. Calculate the relative atomic mass of chlorine.

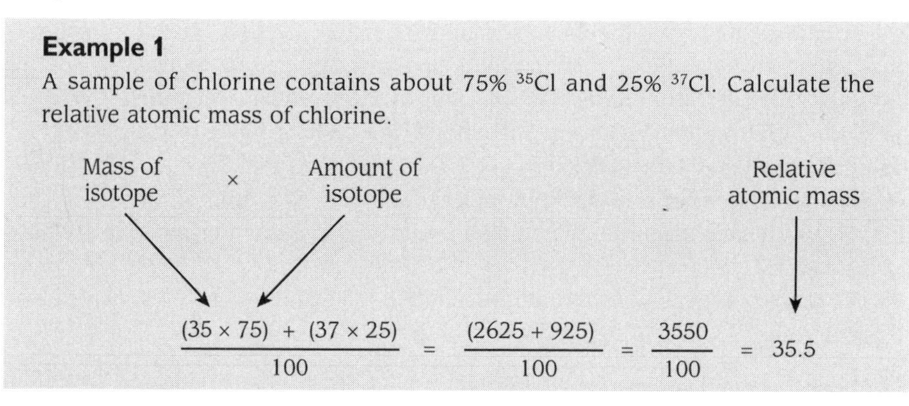

Example 2

A sample of iron contains three isotopes: ^{54}Fe, ^{56}Fe and ^{57}Fe. The relative abundance is 2 : 42 : 1 respectively. Calculate the relative atomic mass.

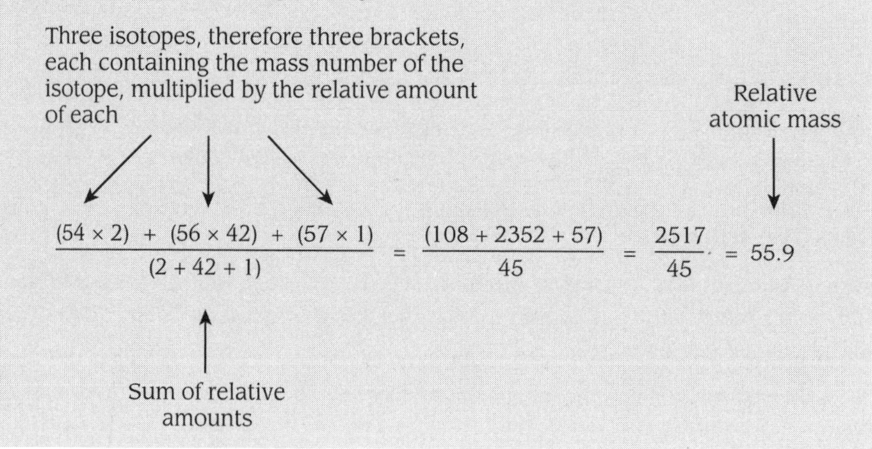

Relative molecular mass and relative formula mass can be calculated by using the relative atomic masses. Relative molecular mass applies to covalent molecules only. Relative formula mass applies to all substances.

Example 1

Calculate the relative molecular mass of (a) carbon dioxide, CO_2, and (b) glucose, $C_6H_{12}O_6$.

(a) $CO_2 = 12 + 16 + 16 = 44$

(b) $C_6H_{12}O_6 = (12 \times 6) + (1 \times 12) + (16 \times 6)$
$= 72 + 12 + 96 = 180$

Example 2

Calculate the relative formula mass of sodium carbonate, Na_2CO_3.

$Na_2CO_3 = (23 \times 2) + 12 + (16 \times 3)$
$= 46 + 12 + 48 = 106$

Example 3

Calculate the relative formula mass of copper sulfate crystals, $CuSO_4{\bullet}5H_2O$.

$CuSO_4{\bullet}5H_2O = 63.5 + 32.1 + (16 \times 4) + 5 \times (1 + 1 + 16)$
$= 63.5 + 32.1 + 64 + 90 = 249.6$

Moles and equations

The mole

The mole is the amount of substance that contains as many single particles as there are atoms in 12 g of the carbon-12 (^{12}C) isotope, and is equal to **Avogadro's constant**, N_A, which has a value of 6.02×10^{23} mol^{-1}.

The mass of 1 mole of molecules of a substance equals the relative molecular mass in grams. This is the molar mass, M_r. The units of molar mass are g mol^{-1}.

The **molar mass** is defined as *'the mass per mole of a substance'* and is given the symbol ***M***.

The amount of substance in moles is given the symbol *'n'*.

Empirical and molecular formulae

It is important to understand the difference between empirical and molecular formulae.

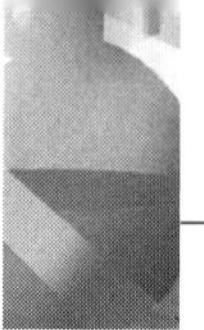

The **empirical formula** is defined as the simplest whole number ratio of atoms of each element in a compound.

The **molecular formula** is defined as the actual number of atoms of each element in a molecule of a compound.

Example

Compound A has a relative molecular mass of 180 and has a composition by mass of C, 40%; H, 6.7%; O, 53.3%. Calculate the empirical formula and the molecular formula.

	C	H	O
Divide the percentage of each element by its own relative atomic mass →	40/12 = 3.3	6.7/1 = 6.7	53.3/16 = 3.3
Divide each by the smallest →	3.3/3.3 = 1	6.7/3.3 = 2	3.33/3.3 = 1

Hence the empirical formula is $C_1H_2O_1 = CH_2O$

Calculate the mass of the empirical formula → $CH_2O = 12 + 2 + 16 = 30$

Deduce how many empirical units are needed to make up the molecular mass → $\frac{\text{Molecular mass}}{\text{Empirical mass}} = \frac{180}{30} = 6$

Therefore the molecular formula is made up of six empirical units, hence the molecular formula = $C_6H_{12}O_6$

Using equations

It is essential that you can write the formulae of a range of common chemicals and write balanced equations. The chemistry of group 2 and group 7 elements (pages 42–48) provides many opportunities to practise this. You should also look back to your GCSE notes and practise writing formulae and equations.

The periodic table can be used to deduce the formula of most chemicals, although there are many exceptions.

Group	1	2	3	4	5	6	7
Number of bonds (valency)	1	2	3	4	3	2	1

If a group 1 element forms a compound with a group 6 element:

- group 1 (e.g. Li) forms one bond
- group 6 (e.g. O) forms two bonds
- two lithium atoms are required for each oxygen atom
- therefore the formula of the compound is Li_2O

If a group 3 element forms a compound with a group 6 element:

- group 3 (e.g. Al) forms three bonds
- group 6 (e.g. O) forms two bonds
- two aluminium ions are required for three oxygen atoms (*each equates to six bonds*)
- therefore the formula of the compound is Al_2O_3

An alternative way of deducing a formula is to use the 'valency cross-over' technique. Using aluminium oxide as an example, follow the simple steps:

Step 1	Write each of the symbols	Al O
Step 2	Write the valency on the top right-hand side of the symbol	Al^3 O^2
Step 3	Crossover the valencies and write them at the bottom right of the other symbol	$Al^3_2 \times O^2_3$
Step 4	You now have the formula of the compound	Al_2 O_3

You are also expected to know the formula of hydrochloric acid (HCl), sulfuric acid (H_2SO_4), nitric acid (HNO_3) and their corresponding salts.

Valencies of elements and groups of elements

1	All group 1 elements, H, Ag and NH_4 (ammonium)	All group 7 elements, OH (hydroxide), NO_3 (nitrate), HCO_3 (hydrogencarbonate)
2	All group 2 elements, Fe, Cu, Zn, Pb, Sn	O, S, SO_4 (sulfate), CO_3 (carbonate)
3	All group 3 elements and Fe	
4	C, Si, Pb, Sn	

You should be able to use the table above to work out most formulae.

More about equations

The importance of being able to provide the correct formulae for substances is that it enables you to write equations for chemical reactions. An equation not only summarises the reactants used and the products obtained, it also indicates the numbers of particles of each substance that are required or produced.

A simple case is the reaction of zinc and sulfur to make zinc sulfide. This is summarised in an equation as:

$$Zn + S \rightarrow ZnS$$

It tells us that one atom of zinc reacts with one atom of sulfur to make one particle of zinc sulfide.

When silver reacts with sulfur:

$$2Ag + S \rightarrow Ag_2S$$

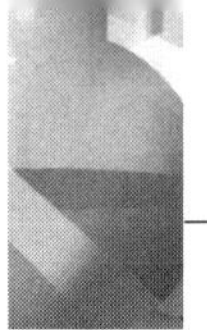

two atoms of silver are required for each atom of sulfur in order to make one particle of silver sulfide.

In any equation all symbols must be balanced and must be the same on each side of the equation.

Calculation of reacting masses, mole concentrations and volumes of gases

Moles from mass

Amount of substance in moles, $n = \frac{\text{Mass of substance (in g)}, m}{\text{Molar mass of substance}, M}$

Such that $n = \frac{m}{M}$

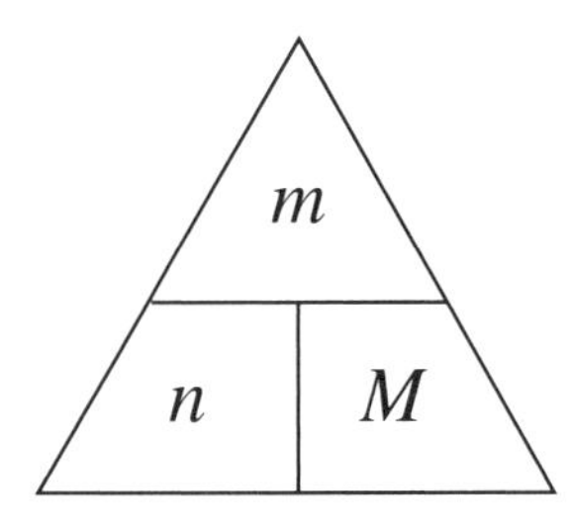

This applies to atoms and to molecules. The examples below illustrate the sort of questions that you may be asked. The triangle can be used to help you rearrange the equation.

Example 1

Calculate the number of moles present in 8.0 g of NaOH.

The information given includes m, the mass (8.0 g) and the formula (NaOH), so it is possible to deduce the molar mass of NaOH. M is calculated by adding together the relative atomic masses of each individual element in the molecule.

Na = 23; O = 16; H = 1, therefore $M = 23 + 16 + 1 = 40$

$n = m/M$, therefore $n = 8.0/40 = 0.2$ mol

Example 2

Calculate the mass of 0.25 moles of $CaCO_3$.

The information given includes n, the number of moles = 0.25 mol

You can use the formula ($CaCO_3$) to deduce the molar mass of $CaCO_3$. The M is calculated by adding together the relative atomic masses of each individual element in the molecule.

Ca = 40.1; C = 12; O = 16 (but there are three O atoms, therefore $3 \times 16 = 48$) = 100.1

$n = m/M$, which can be rearranged to give $nM = m$

$0.25 \times 100.1 = m$, therefore m is 25.025 g

It would be unusual to quote an answer to five significant figures, and you may be asked to quote the answer to three or four significant figures, in which case m is 25.0 g (to three significant figures), or 25.03 (to four significant figures).

Mole calculations for gases

Chemical reactions also take place in the gas phase, and it is therefore important to be able to calculate amounts of chemicals present in gases. Avogadro deduced that all gases, under the same conditions of temperature and pressure, occupy the same volume. At room temperature and pressure the volume of one mole of a gas is equal to 24 dm^3.

It follows that the number of moles of a gas can be calculated using:

Volume of gas in dm^3:

Amount of gas in mols: $n = \dfrac{V \text{ (in dm}^3)}{24}$

Volume of gas in cm^3:

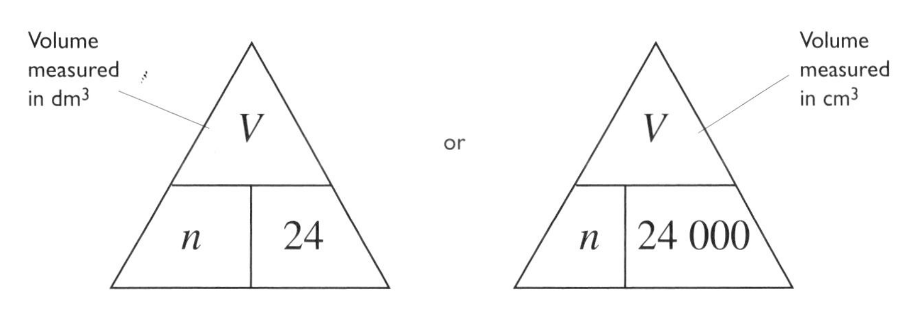

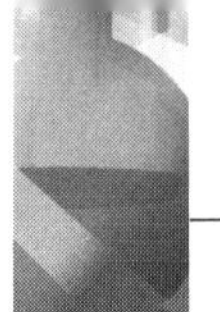

Example

Calculate the number of moles present in $120\,cm^3$ of hydrogen at room temperature and pressure.

Use $n = V/24\,000$ (since the volume of gas has been given in cm^3):

$$n = 120/24\,000$$
$$= 0.005 = 5 \times 10^{-3}$$

Mole calculations for solutions

Many chemical reactions are carried out in solution (often dissolved in water). The amount of chemical present is best described by using the concentration of the solution, c, in $mol\,dm^{-3}$ and the volume of the solution, V.

The units of concentration are nearly always $mol\,dm^{-3}$, and it follows that V, the volume of the solution, is measured in dm^3.

The amount in mol for solutions can be calculated using $n = c \times V$.

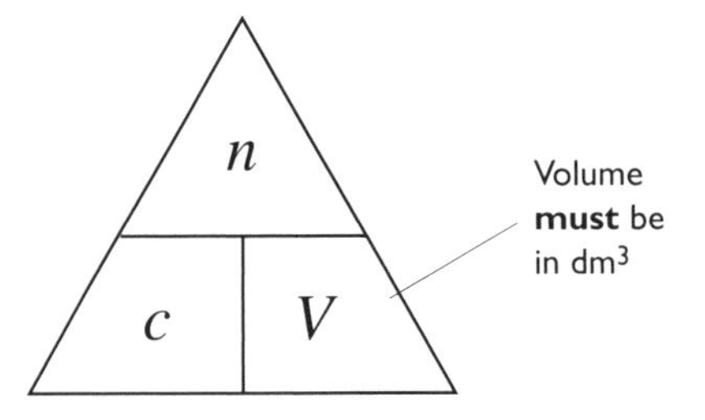

If V is given in cm^3, then use:

$$n = c \times \frac{V\ (\text{in cm}^3)}{1000}$$

The examples below illustrate the sort of questions that you may be asked.

Example 1

Calculate the number of moles present in $250\,cm^3$ of $0.50\,mol\,dm^{-3}$ solution.

The concentration, c, is given as $0.50\,mol\,dm^{-3}$, and the volume, V, is $250\,cm^3$. The volume must be converted into dm^3, i.e. $V = 250/1000 = 0.25\,dm^3$.

$$n = c \times V$$
$$= 0.5 \times 0.25 = 0.125\ mol$$

Example 2

Calculate the concentration of a sodium hydroxide (NaOH) solution when 4.0 g NaOH is dissolved in 250 cm^3 of water.

This is slightly more complicated, and you first have to calculate the number of moles of NaOH using the mass (4.0 g) and the formula, NaOH, to deduce the molar mass (23 + 16 + 1 = 40), such that:

$n = m/M = 4/40 = 0.1$ mol

The concentration can then be calculated, using:

$c = n/V = 0.1/(250/1000) = 0.1/0.25 = 0.4 \text{ mol dm}^{-3}$

Using equations

Mole calculations appear on almost every Unit F321 paper. You should be able to:

- write and balance full and ionic equations
- calculate reacting masses
- calculate reacting gas volumes
- calculate reacting volumes of solutions
- calculate concentrations from titrations

The calculation below shows how reacting masses, reacting gas volumes and reacting volumes of solutions could all be tested in a single equation.

Example

Zinc reacts with dilute hydrochloric acid to produce zinc chloride and hydrogen gas. Write a balanced equation for this reaction and calculate the mass of zinc required to react exactly with 50 cm^3 of a 0.10 mol dm^{-3} solution of HCl(aq). Deduce the volume, in cm^3, of hydrogen gas produced in this reaction.

Equation:	Zn	+	2HCl	→	$ZnCl_2$	+	H_2

The number in front of each formula in the balanced equation tells us the number of moles used and gives us the ratio of the reacting moles, i.e. the mole ratio:

Equation:	Zn	+	2HCl	→	$ZnCl_2$	+	H_2
Mole ratio:	1 mol	:	2 mol	:	1 mol	:	1 mol

We can calculate n for HCl, since we are given c and V.

$n = 0.1 \times 50/1000 = 0.0005 = 5 \times 10^{-3}$ mol.

Once we know the moles of one chemical in the equation, we can deduce the number of moles of all the other chemicals in the equation by using the mole ratio.

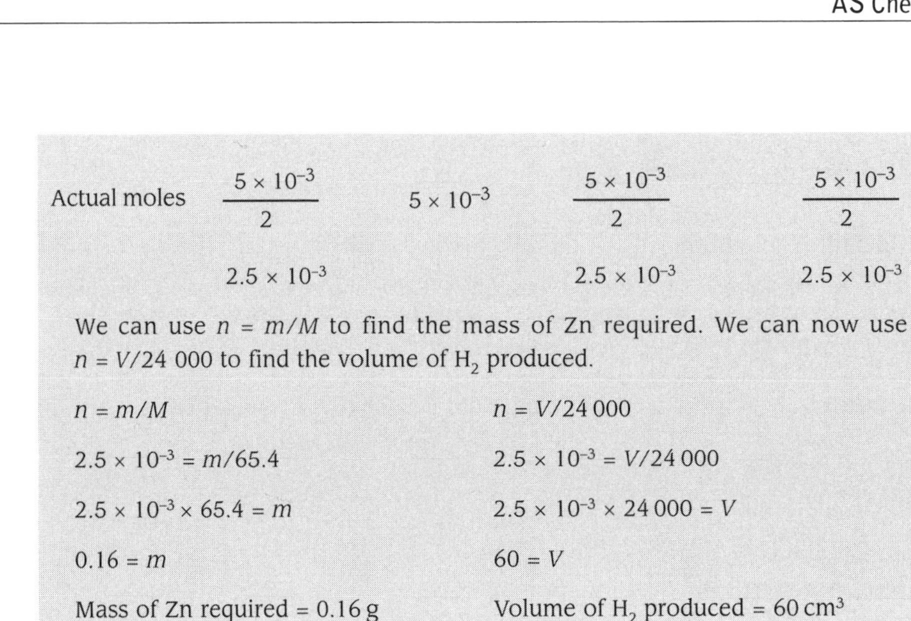

Actual moles	$\frac{5 \times 10^{-3}}{2}$	5×10^{-3}	$\frac{5 \times 10^{-3}}{2}$	$\frac{5 \times 10^{-3}}{2}$
	2.5×10^{-3}		2.5×10^{-3}	2.5×10^{-3}

We can use $n = m/M$ to find the mass of Zn required. We can now use $n = V/24\,000$ to find the volume of H_2 produced.

$n = m/M$	$n = V/24\,000$
$2.5 \times 10^{-3} = m/65.4$	$2.5 \times 10^{-3} = V/24\,000$
$2.5 \times 10^{-3} \times 65.4 = m$	$2.5 \times 10^{-3} \times 24\,000 = V$
$0.16 = m$	$60 = V$
Mass of Zn required = 0.16 g	Volume of H_2 produced = 60 cm^3

Acids

Acids and bases

Acids are defined as proton donors, and bases can be described as proton acceptors.

The proton, H^+, released by an acid can only occur in aqueous solution. Pure hydrogen chloride, for example, is a covalent gas, and it is only when it comes into contact with water that it can release a proton.

$HCl(g) + H_2O(l) \rightarrow H_3O^+(aq) + Cl^-(aq)$

$H_3O^+(aq)$ is often written as $H^+(aq)$, and the equation shown above is often given as:

$HCl(aq) \rightarrow H^+(aq) + Cl^-(aq)$

You are expected to know the formulae of the three common acids below, and you will not be given these formulae in an examination.

- hydrochloric acid HCl
- sulfuric acid H_2SO_4
- nitric acid HNO_3

Acids have a pH below 7, and the stronger acids have lower pHs.

Bases are defined as proton (H^+) acceptors. An alkali is defined as a soluble base that can release hydroxide ions, OH^-, when in aqueous solution.

Common bases include metal oxides (e.g. Na_2O, MgO, CuO), metal hydroxides (e.g. $NaOH$, KOH, $Cu(OH)_2$) and ammonium hydroxide (NH_4OH).

Common alkalis include group 1 hydroxides such as LiOH, NaOH and some group 2 hydroxides, such as $Ca(OH)_2$.

Salts

A salt is defined as a substance that is formed when one or more H^+ ions of an acid are replaced by a metal ion or an ammonium ion, NH_4^+.

Salts are formed when acids react with:

- a metal
- a carbonate
- a base
- an alkali

Acid and metal

Balanced equation: $2HCl(aq) + Zn(s) \rightarrow ZnCl_2(aq) + H_2(g)$

Ionic equation: $2H^+(aq) + Zn(s) \rightarrow Zn^{2+}(aq) + H_2(g)$

Acid and carbonate

Balanced equation: $2HCl(aq) + Na_2CO_3(aq) \rightarrow 2NaCl(aq) + H_2O(l) + CO_2(g)$

Ionic equation: $2H^+(aq) + CO_3^{2-}(aq) \rightarrow H_2O(l) + CO_2(g)$

Acid and base

Balanced equation: $2HCl(aq) + MgO(s) \rightarrow MgCl_2(aq) + H_2O(l)$

Ionic equation: $2H^+(aq) + MgO(s) \rightarrow Mg^{2+}(aq) + H_2O(l)$

Acid and alkali

Balanced equation: $HCl(aq) + NaOH(aq) \rightarrow NaCl(aq) + H_2O(l)$

Ionic equation: $H^+(aq) + OH^-(aq) \rightarrow H_2O(l)$

Bases such as ammonia can react with acids to produce a salt. Ammonium sulfate $(NH_4)_2SO_4$ is used as a fertiliser, and is manufactured by reacting ammonia with sulfuric acid.

Balanced equation: $2NH_3(aq) + H_2SO_4(aq) \rightarrow (NH_4)_2SO_4(aq)$

Ionic equation: $2NH_3(aq) + 2H^+(aq) \rightarrow 2NH_4^+(aq)$

Salts such as copper sulfate can exist as either the anhydrous salt, $CuSO_4$, or as the hydrated salt, $CuSO_4 \cdot 5H_2O$. Anhydrous copper sulfate is a white powder, while hydrated copper sulfate exists as shiny blue crystals. They appear crystalline due to the water of crystallisation.

You are expected to be able to calculate the formula of a hydrated salt from percentage composition data or from experimental data. It is quite likely that one of the practical tasks will be set around this area of the specification.

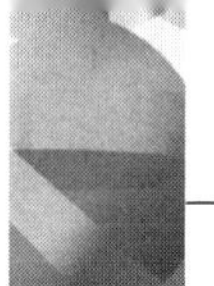

Example

2.86 g of hydrated sodium carbonate, $Na_2CO_3 \cdot xH_2O$ was heated to constant mass. The mass of the remaining anhydrous sodium carbonate, Na_2CO_3, was 1.06 g. Use these data to calculate the value of x.

mass of water in $Na_2CO_3 \cdot xH_2O$ = 2.86 – 1.06 = 1.80 g

molar mass of water = 1 + 1 + 16 =18

moles of water = 1.80/18 = 0.1 mol

mass of anhydrous Na_2CO_3 = 1.06 g

molar mass of Na_2CO_3 = 23 + 23 + 12 + (3 × 16) = 106

moles of anhydrous Na_2CO_3 = 1.06/106 = 0.01 mol

	Na_2CO_3	H_2O
mols	0.01 :	0.1
simplest ratio	$\frac{0.01}{0.01}$:	$\frac{0.1}{0.01}$
	= 1 :	10

The formula of the hydrated crystals is $Na_2CO_3 \cdot 10H_2O$, such that the value of x is 10.

You will also be expected to carry out acid–base titrations in the laboratory, and then to undertake structured calculations. A titration is a process whereby a precise volume of one solution is added to another solution, until the exact volume required to complete the reaction has been found. With care it is possible to obtain reproducible results accurate to within 0.10 cm^3.

Example

It was found that 18.60 cm^3 HCl(aq) exactly neutralised 25 cm^3 0.100 mol dm^{-3} NaOH(aq). Calculate the concentration of the HCl(aq) solution.

Step 1 is to work out how many moles of sodium hydroxide were neutralised. This is possible, as it is the solution for which you know both the concentration used, c, and the volume, V ($n = cV = 0.100 \times 0.0250$).

The concentration of the sodium hydroxide is 0.100 mol dm^3 of solution. Since 25.0 cm^3 was used, this would contain 0.100 × (25.0/1000) = 0.00250 mol

Step 2 is to refer to the balanced equation to see how many moles of hydrochloric acid would be needed to react with this number of moles of sodium hydroxide.

The equation for the reaction is:

	NaOH	+	HCl	→	NaCl	+	H_2O
The mole ratio is	1	:	1	:	1	:	1

and therefore 1 mole of NaOH reacts with 1 mole of HCl.

Since we used 0.00250 moles of NaOH, it must have required 0.00250 moles of HCl to react completely. Since 18.60 cm^3 of HCl were added from the burette, it must follow that 18.60 cm^3 contained 0.00250 moles of acid.

Step 3 is to convert the information obtained about the hydrochloric acid into its concentration in mol dm^{-3}.

$$c = \frac{n}{V} = \frac{0.00250}{18.6/1000} = \frac{0.00250}{0.0186} = 0.134 \text{ mol dm}^{-3}$$

So concentration of HCl was 0.134 mol dm^{-3}.

Redox

Redox reactions are reactions in which electrons are transferred from one substance to another. There are numerous everyday examples of redox reactions, including the combustion of fuels and the rusting of iron. Oxidation was originally defined as the gaining of oxygen. For example, when zinc reacts with oxygen to form zinc oxide, the zinc has clearly been oxidised.

$Zn(s) + \frac{1}{2}O_2(g) \rightarrow ZnO(s)$

A closer inspection of this reaction shows that the zinc atom has been converted to a zinc ion, and in the process has lost two electrons.

$Zn \rightarrow Zn^{2+} + 2e^-$

The definition of oxidation therefore has been extended, so that a species is said to be oxidised if it loses electrons. The converse is true for reduction. One way of remembering this is shown below.

OILRIG

Oxidation Is Loss   Reduction Is Gain

Oxidation number

The oxidation number is a convenient way of quickly identifying whether or not a substance has undergone either oxidation or reduction. In order to work out the oxidation number, you must first learn a few simple rules.

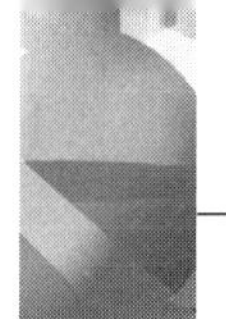

Rule	Example
All elements in their natural state have an oxidation number zero	H_2; oxidation number of H = 0
The oxidation numbers of the atoms of any molecule add up to zero	H_2O; sum of oxidation numbers = 0
The oxidation numbers of the components of any ion add up to the charge of the ion	SO_4^{2-}; sum of oxidation numbers = –2

There are certain elements whose oxidation numbers never change, but some other elements have variable oxidation numbers, and so these have to be deduced.

When calculating the oxidation numbers of elements in either a molecule or an ion, you should apply the following order of priority.

(1) Group 1, 2 and 3 elements are always +1, +2 and +3 respectively.
(2) Fluorine is always –1.
(3) Hydrogen is usually +1.
(4) Oxygen is usually –2.
(5) Chlorine is usually –1.

By applying these in sequence it is possible to deduce any other oxidation number.

Example 1

Deduce the oxidation number of the chlorine in sodium hypochlorite (NaClO).

The sum of the oxidation numbers in NaClO must add up to zero.

In order of priority, sodium comes first and must be +1; oxygen comes second and is –2. In order for the oxidation numbers to add up to zero, the oxidation number of Cl must be +1.

Example 2

Deduce the oxidation number of Cl in the chlorate ion (ClO_3^-).

The oxidation numbers must add up to the charge on the ion, i.e. they must add up to –1.

In order of priority, oxygen comes first and is –2, but there are three oxygens, and hence the total is –6. In order for the oxidation numbers to add up to the charge of the ion (–1) the chlorine must be +5.

When magnesium reacts with steam, magnesium oxide and hydrogen are formed:

$$Mg(s) + H_2O(g) \rightarrow MgO(s) + H_2(g)$$

It is easy to see that magnesium has been oxidised because it has gained oxygen, and that water has been reduced because it has lost oxygen. When this reaction is investigated using oxidation numbers, it shows:

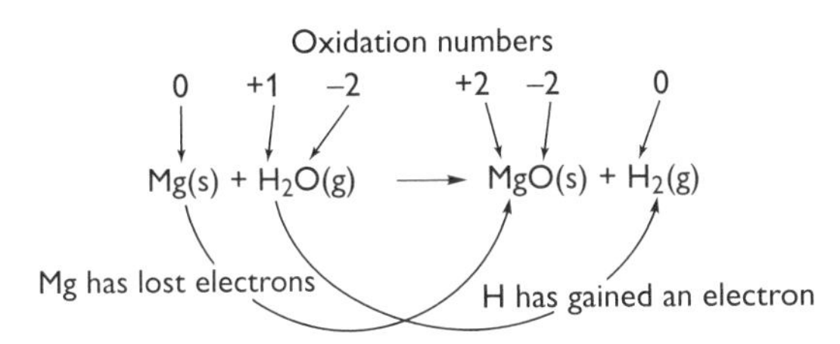

From this it is clear that an increase in oxidation number is *oxidation*, while a decrease in oxidation number is *reduction*.

Electron transfer can be shown by using half-equations, such that:

$Mg \rightarrow Mg^{2+} + 2e^-$ (*loss* of electrons is oxidation)

$2H^+ + 2e^- \rightarrow H_2$ (*gain* of electrons is reduction)

A number of elements can have more than one valency. Iron reacts with chlorine and it is possible to form two different chlorides, $FeCl_2$ and $FeCl_3$, so that Fe can have either a valency of two or a valency of three. Since there is a possibility of confusion, the name given to the compounds must make the valency absolutely clear. Iron(II) chloride ($FeCl_2$) is so named to indicate that it is a compound containing iron with a valency of two, while iron(III) chloride makes it clear that we are referring to $FeCl_3$.

Redox reactions

Metals generally react by losing electrons (oxidation is loss), causing the oxidation number to increase.

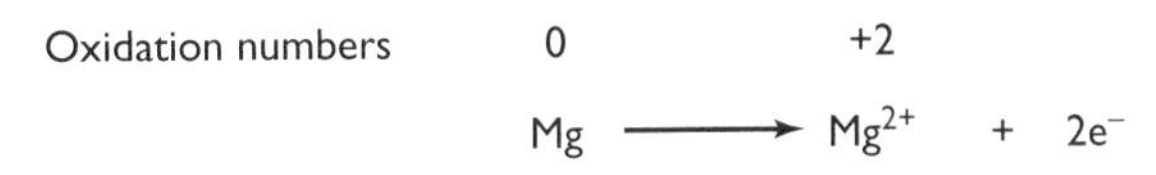

Non-metals generally react by gaining electrons (reduction is gain), causing the oxidation number to decrease.

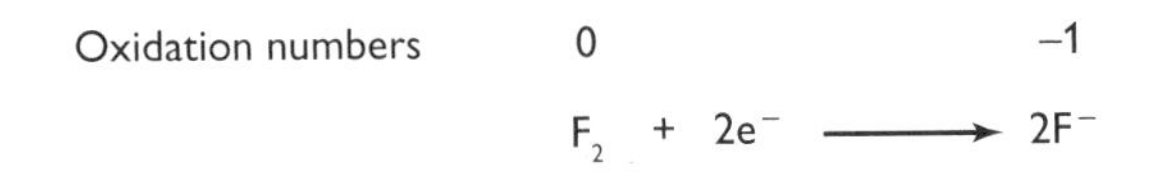

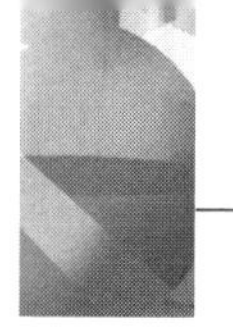

When a metal reacts with either HCl(aq) or H_2SO_4(aq), the metal is oxidised.

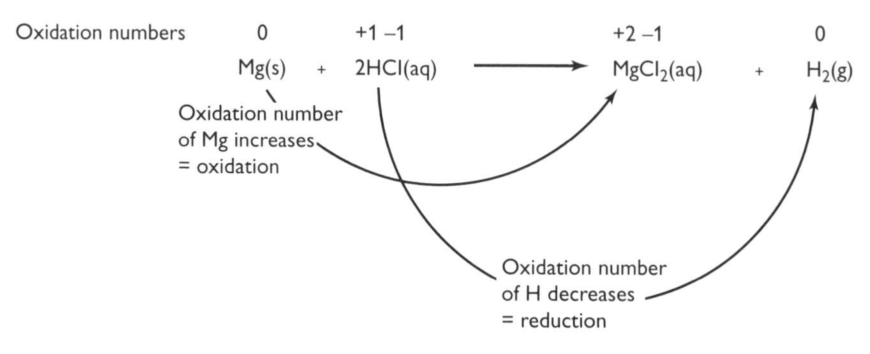

You should be able to use oxidation numbers to determine what has been oxidised or reduced in unfamiliar reactions, such as:

$$4HCl + MnO_2 \rightarrow MnCl_2 + Cl_2 + 2H_2O$$

Work out the oxidation number of each element on both sides of the equation. Identify the element whose oxidation number has *increased* (oxidation) and the element whose oxidation number has *decreased* (reduction).

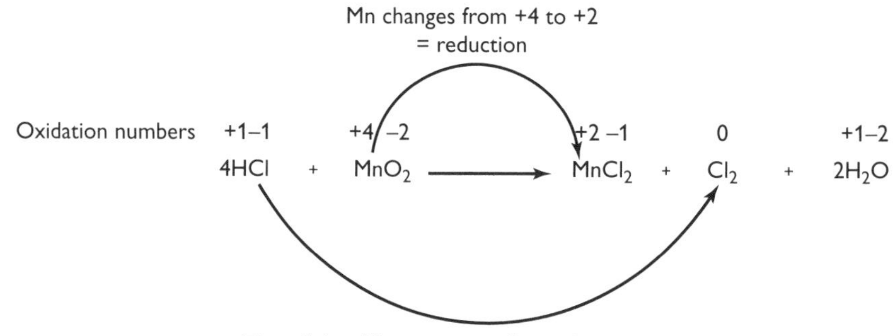

Electrons, bonding and structure

Electronic structure of atoms

Ionisation energy provides evidence for the existence of electron shells and sub-shells.

You should be able to define the **first ionisation energy**. This is the energy required to remove one electron from each atom in 1 mole of gaseous atoms.

The first ionisation energy can be represented by the equation:

$$X(g) \rightarrow X^+(g) + e^-$$

It is important to include the state symbols.

For elements that have more than one electron, it is possible to remove electrons one by one from the atom.

The second ionisation energy is defined as the energy required to remove one electron from each ion with 1+ charge in 1 mole of gaseous ions. The second ionisation energy can be represented by the equation:

$X^{+}(g) \rightarrow X^{2+}(g) + e^{-}$

The second ionisation energy results in the formation of a 2+ ion and starts with a 1+ ion. The third ionisation energy will form a 3+ ion and start with a 2+ ion. It follows that the eighth ionisation energy is:

$X^{7+}(g) \rightarrow X^{8+}(g) + e^{-}$

Trends in ionisation energies

There are three factors that influence ionisation energy:

- the distance of the outermost electron from the nucleus (atomic radius)
- electron shielding
- nuclear charge

As we go across a period, the atomic radii decrease and the main shell shielding remains the same. This should make it more difficult to remove an electron. The nuclear charge also increases across a period, making it more difficult to remove an electron. Therefore, ionisation energy increases across a period.

As we go down a group, atomic radii and shielding both increase, and this should make it easier to remove an electron, but as we go down a group the nuclear charge also increases, which should make it harder to remove an electron. Increases in atomic radii and shielding outweigh nuclear charge, so that ionisation energy decreases down a group.

It is possible to remove each electron, one by one, from an atom, and to measure the size of each successive ionisation energy. When the successive energies are plotted, the graph provides evidence for the existence of shells. Chlorine has 17 electrons, which we know to be arranged as 2,8,7, in three electron shells. The evidence for this is shown in a plot of successive ionisation energies.

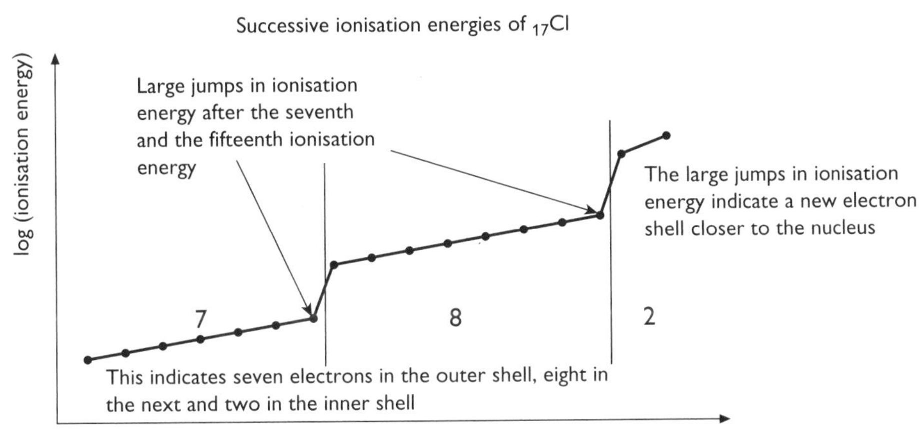

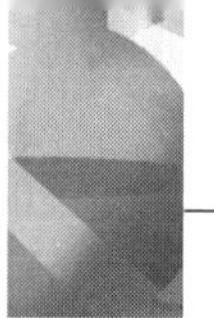

Plotting successive ionisation energies confirms what you learnt at GCSE. It shows clearly that the first shell contains a maximum of two electrons and the second shell contains a maximum of eight electrons. There is further experimental evidence to suggest that each shell is made of smaller sub-shells.

By studying the first ionisation energy of the first 20 elements, evidence for the existence of sub-shells was obtained.

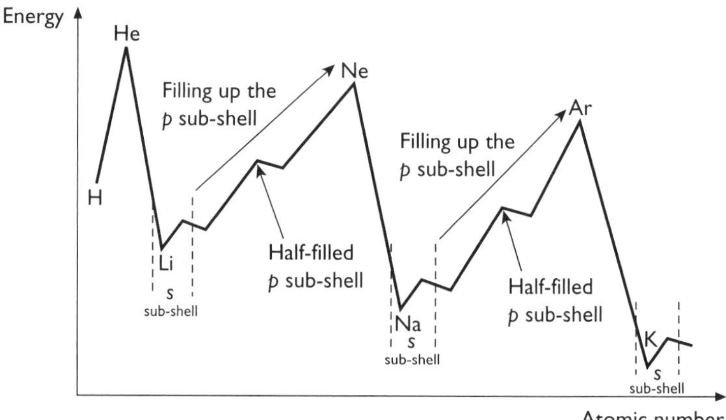

The trends in ionisation energy suggested that each group 1 element would have a low first ionisation energy and that this ionisation energy would increase across the period to reach a maximum at group 8 (the noble gases). There is a gradual increase in ionisation energies across a period, but several small peaks and troughs exist. These peaks and troughs are repeated in each period. They provide evidence for the existence of sub-shells (*s*, *p* and *d*). There is a periodic variation throughout all of the elements and there is evidence for several shells and sub-shells.

Shell	**Sub-shells**				**Total number of electrons**			
First	1*s*				2			
Second	2*s*	2*p*			2	6		
Third	3*s*	3*p*	3*d*		2	6	10	
Fourth	4*s*	4*p*	4*d*	4*f*	2	6	10	14

We now know that the sub-shells are made up of orbitals.

Orbitals

The concept of an orbital is difficult. If you are asked in an exam to define an orbital, the simplest definition is 'a volume in space where there is a high probability of finding an electron'. An orbital can hold up to a maximum of two electrons, each spinning in opposite directions. You should be able to describe, with the aid of a diagram, the shape of *s*- and *p*-orbitals.

There are three *p*-orbitals, one along each of the *x*-, *y*- and *z*-axes.

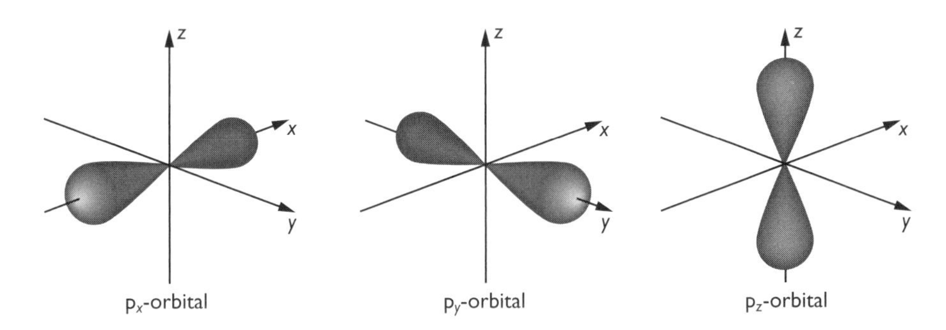

The sequence in which electrons fill the orbitals is shown in the energy diagram below.

Remember that:

- within an orbital, the electrons must have opposite spins
- the lowest energy level is occupied first
- orbitals at the same energy level are occupied singly before pairing of electrons

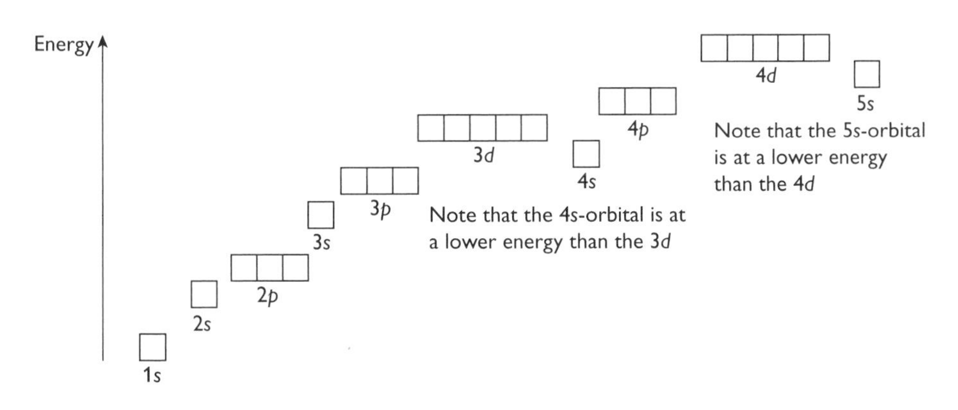

Electronic configuration

You should be able to write the full electronic configuration of the first 36 elements in the following manner:

- ${}_{17}$Cl is $1s^2\ 2s^2\ 2p^6\ 3s^2\ 3p^5$ or [Ne] $3s^2\ 3p^5$
- ${}_{26}$Fe is $1s^2\ 2s^2\ 2p^6\ 3s^2\ 3p^6\ 3d^6\ 4s^2$ or [Ar] $3d^6\ 4s^2$

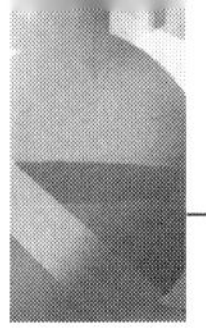

You should also be able to write the full electronic configuration of the ions of the first 36 elements in the following manner:

- $_{17}Cl^-$ is $1s^2\ 2s^2\ 2p^6\ 3s^2\ 3p^6$ or [Ne] $3s^2\ 3p^6$
- $_{26}Fe^{2+}$ is $1s^2\ 2s^2\ 2p^6\ 3s^2\ 3p^6\ 3d^6$ or [Ar] $3d^6$

It is a common mistake to write $_{26}Fe$ as $1s^2\ 2s^2\ 2p^6\ 3s^2\ 3p^6\ 4s^2\ 3d^4$ and then remove the $3d$ electrons first, such that $_{26}Fe^{2+}$ is written as $1s^2\ 2s^2\ 2p^6\ 3s^2\ 3p^6\ 4s^2\ 3d^2$. This is *not* the case and it will cost you marks in the exam. It is worth remembering that the $3d$ electrons are part of an inner shell and the outer shell ($4s$) electrons will always be lost first.

The electronic configuration can also give a clue as to relative stability. Comparing the electronic configuration of Fe^{2+} and Fe^{3+}:

- $_{26}Fe^{2+}$ is $1s^2\ 2s^2\ 2p^6\ 3s^2\ 3p^6\ 3d^6\ 4s^2$ or [Ar] $3d^6$
- $_{26}Fe^{3+}$ is $1s^2\ 2s^2\ 2p^6\ 3s^2\ 3p^6\ 3d^5\ 4s^2$ or [Ar] $3d^5$

This shows that Fe^{3+} has a half-filled *d* shell ($3d^5$), while one of the *d*-orbitals in Fe^{2+} is occupied by two electrons and hence is not as stable as Fe^{3+}.

The full electronic configuration allows you to classify elements into *s*-, *p*- and *d*-block elements and to identify their positions in the periodic table. For example:

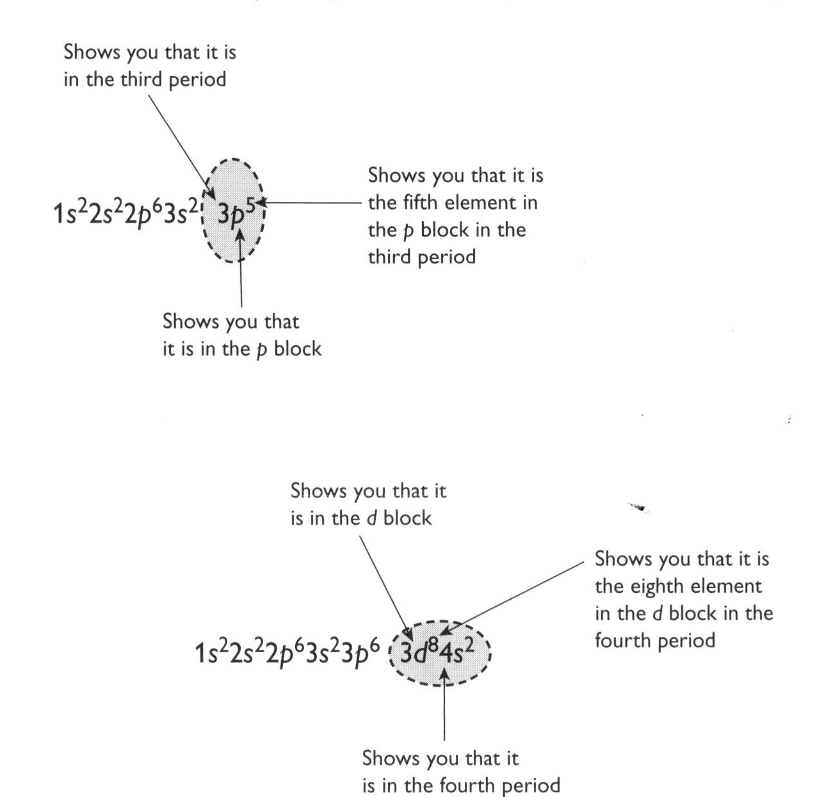

Chemical bonding and structure

Types of bond

There are three main types of bond: **ionic**, **covalent** and **metallic**. The table below summarises each type.

	Ionic	Covalent	Metallic
Definition	An ionic bond is the electrostatic attraction between oppositely charged ions	A covalent bond is a shared pair of electrons	A metallic bond is the electrostatic attraction between positive metal ions in the lattice, and delocalised electrons (often referred to as a 'sea of free electrons')
Formation	Formed by electron transfer from metal atom (X) to non-metal atom (Y) to produce oppositely charged ions X^+ and Y^-	Formed when electrons are shared rather than transferred	The positive ions occupy fixed positions in a lattice and the delocalised electrons can move freely throughout the lattice
Direction	An ionic bond is directional, acting between adjacent ions	A covalent bond is directional, acting solely between the two atoms involved in the bond	A metallic bond is non-directional, because the delocalised electrons can move anywhere in the lattice
Examples	NaCl, MgO	H_2, CH_4	Cu, Na
Melting and boiling points	High melting point and boiling point due to strong electrostatic forces between ions in the solid lattice	Low melting point and boiling point — the simple covalent molecules are held together by weak forces between molecules and hence little energy is needed to break the weak intermolecular forces	High melting point and boiling point due to strong metallic bonds between positive ions and the delocalised electrons in the lattice
Conductivity	Non-conductor of electricity in solid state, but conducts when melted or dissolved in water because the ionic lattice breaks down and ions are free to move as mobile charge carriers	Non-conductors of electricity — no free or mobile charged particles	Good thermal and electrical conductors due to mobile, delocalised electrons which conduct heat and electricity, even in the solid state
Solubility	The ionic lattice usually dissolves in polar solvents (e.g. water) Polar water molecules attract ions in lattice and surround each ion (hydration)	Simple molecular structures soluble in non-polar solvents (e.g. hexane), but usually insoluble in water	Insoluble in polar and non-polar solvents. Some metals (groups 1 and 2) react with water

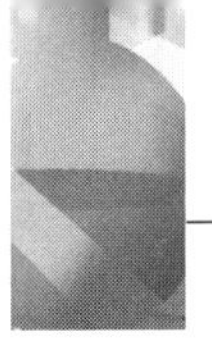

In addition to covalent bonding, **dative covalent** or **coordinate** bonds exist. These are also are the result of two shared electrons, but in this case only one of the atoms supplies *both* shared electrons.

Dot-and-cross diagrams

Dot-and-cross diagrams are a simple visual way to illustrate both ionic and covalent bonding.

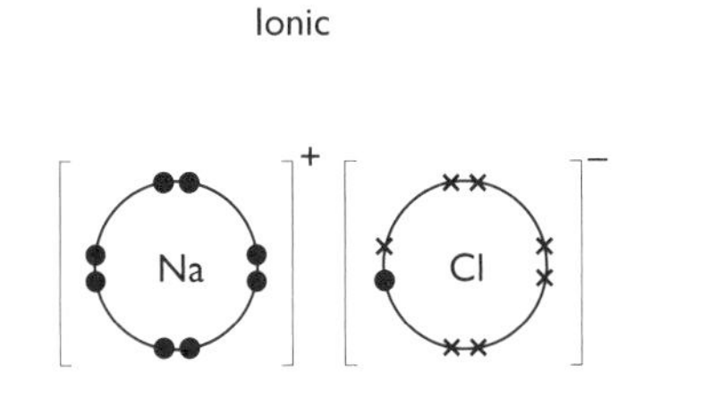

Must use a 'dot' to show transfer of an electron

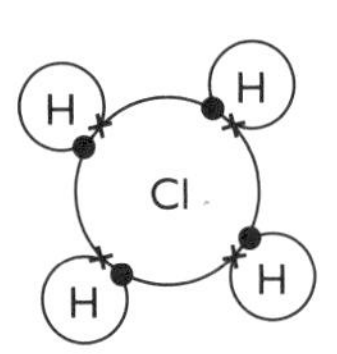

Each bond is made up of a 'dot' and a 'cross'

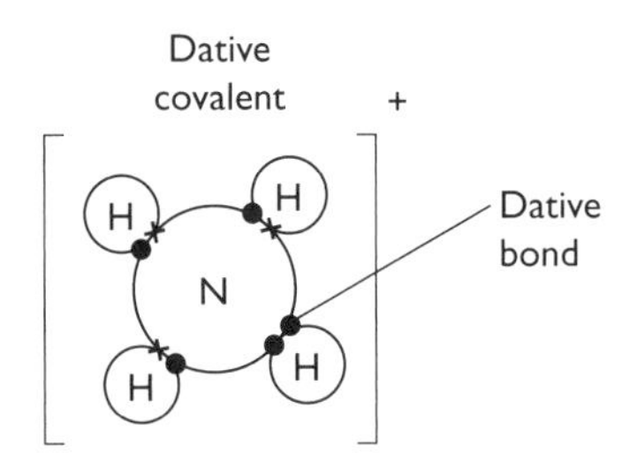

The dative bond is made up of two 'dots'

Shapes of molecules

The **electron-pair repulsion theory** states that the shape of a covalent molecule is determined by the number and nature of electron pairs around the central atom.

- Electron pairs repel one another so that they are as far apart as possible.
- The shape depends upon the number and type of electron pairs surrounding the central atom.
- Lone pairs of electrons are more 'repelling' than bonded pairs of electrons.

You should be able to draw a dot-and-cross diagram of a molecule and use it to determine the number and type of electron pairs around the central atom. You can then use the following table to predict the shape and the bond angle.

Number of bonded pairs of electrons	Number of lone pairs of electrons	Shape	Approximate bond angle
2	0	Linear	180°
3	0	Trigonal planar	120°
4	0	Tetrahedral	109.5°
5	0	Trigonal-bipyramidal	90° and 120°
6	0	Octahedral	90°
3	1	Pyramidal	107°
2	2	Angular	104°

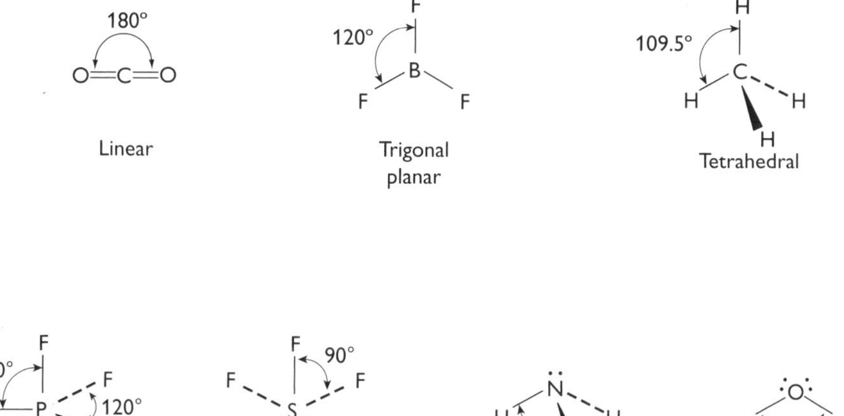

Bond polarity and polarisation

Ionic and covalent bonding are extremes — there is a whole range of intermediate degrees of bonding in between. Many bonds are described as:

- essentially covalent with some ionic character, due to differences in **electronegativity** — bond polarity; or
- essentially ionic with some covalent character, due to differences in **charge density** — polarisation

If a covalent bond is formed between two different elements it is highly likely that each of the elements will attract the covalent bonded pair of electrons unequally.

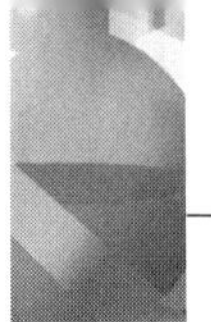

Electronegativity

This is the attraction of an atom, in a molecule, for the pair of electrons in a covalent bond. Electronegativity increases across a period but decreases down a group, so that, for example, fluorine is the most electronegative element.

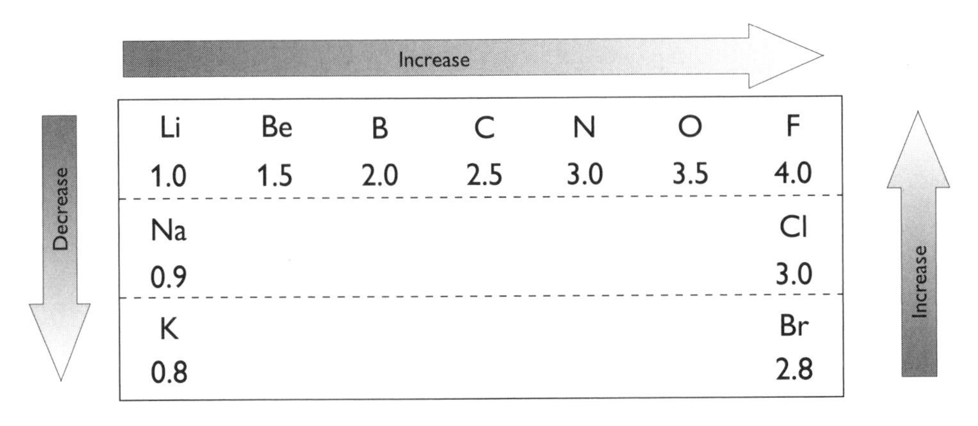

Li	Be	B	C	N	O	F
1.0	1.5	2.0	2.5	3.0	3.5	4.0
Na						Cl
0.9						3.0
K						Br
0.8						2.8

In molecules like hydrogen chloride (HCl), the two electrons in the covalent bond are shared unequally. Chlorine has a higher electronegativity than hydrogen, so that the two shared electrons are pulled towards the chlorine atom, resulting in the formation of a permanent dipole. The bond in HCl consists of two shared electrons (essentially covalent), but also contains $\delta+$ and $\delta-$ charges (hence the ionic character).

- The greater the *difference* between electronegativities, the greater is the *ionic* character of the bond.
- The greater the *similarity* in electronegativities, the greater is the *covalent* character of the bond.

Polarity

If a compound consists of two or more different non-metals bonded together, the bonds are most likely to be essentially covalent with some ionic character. Generally, a compound made up of two or more different non-metals has polar molecules unless the molecules are symmetrical, in which case any dipoles cancel out.

Common examples include hydrogen chloride, water and ammonia.

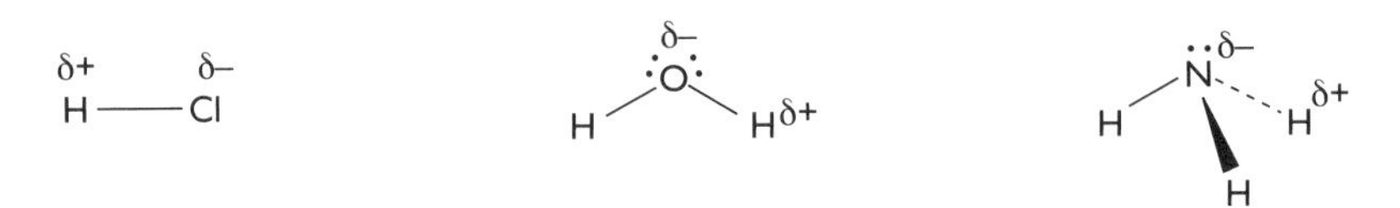

Dipoles usually cancel out if the molecules have any of the following shapes: linear, trigonal planar, tetrahedral, trigonal-bipyramidal or octahedral. Molecules with these shapes tend to be non-polar.

Intermolecular forces

The three main categories of bonds (ionic, covalent and metallic) described above are all strong: 200–600 kJ mol^{-1} are needed to break them. There are three other types of bonds, known collectively as intermolecular forces. By contrast, these are weak bonds needing 2–40 kJ mol^{-1} to break them. The intermolecular forces are:

- van der Waals forces
- permanent dipole–dipole interactions
- hydrogen bonds

Van der Waals forces (induced dipole–dipole interactions) are the weakest of the forces and act between all particles, polar or non-polar. They are caused by the movement of electrons. This movement results in an oscillating or instantaneous dipole, which in turn induces a dipole on neighbouring molecules. The attraction between these induced dipoles produces weak intermolecular interactions called van der Waals forces. The strength of the van der Waals forces depends on the number of electrons in the molecule. The greater the number of electrons in an atom or molecule, the greater are the van der Waals forces.

If you are asked to explain or define a van der Waals force, there are three key features that you must include:

- the movement of electrons generates an instantaneous dipole
- the instantaneous dipole induces another dipole in neighbouring atoms or molecules
- the attraction between the temporary induced dipoles results in the van der Waals forces

Permanent dipole–dipole interactions are usually found between polar molecules that are essentially covalent but have some ionic character.

They are weak intermolecular forces between the permanent dipoles of adjacent molecules.

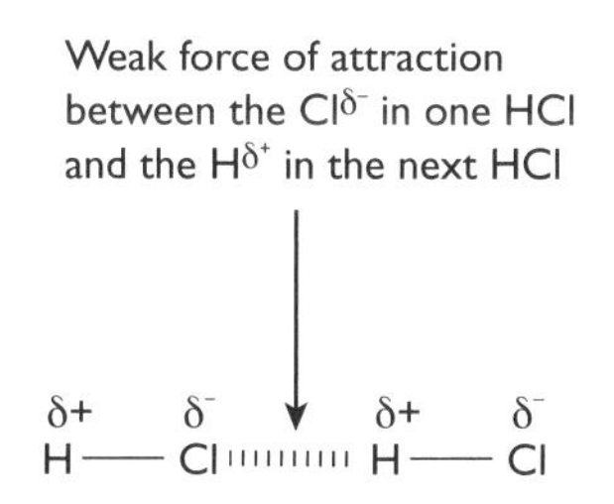

Hydrogen bonds exist between molecules that contain hydrogen that is bonded to either nitrogen, oxygen or fluorine. They are comparatively strong permanent dipole–dipole interactions, involving a lone pair of electrons on either the nitrogen, oxygen or fluorine atom. Hydrogen bonds exist between molecules of ammonia, water and hydrogen fluoride. They are also found in alcohols.

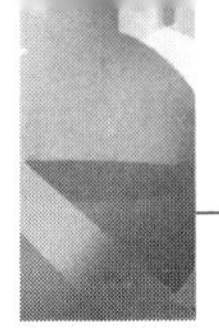

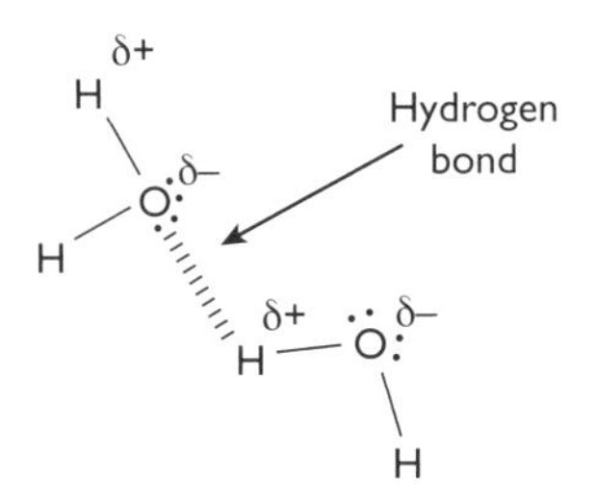

Special properties of water arise from hydrogen bonding:

- The solid (ice) is less dense than the liquid (water) because the hydrogen bonds in ice hold the H_2O molecules further apart, creating a more open lattice structure.
- Water has a relatively high melting point and boiling point, due to the additional energy required to break the hydrogen bonds.

Bonding and physical properties

Giant ionic and giant metallic lattices and their properties are covered in the table on page 33.

Giant covalent structures

These are also known as giant molecular structures. They are held together by strong covalent bonds between atoms. Some important properties are:

- high melting points and boiling points
- non-conductors of electricity, because there are no mobile or free charged particles (except in graphite)
- insoluble in polar and non-polar solvents, due to the strong covalent bonds in the lattice

Diamond and graphite both have giant molecular structures. Diamond is a giant structure, with each carbon atom bonded to four other carbon atoms. Graphite is also giant structure, with each carbon atom bonded to three other carbon atoms.

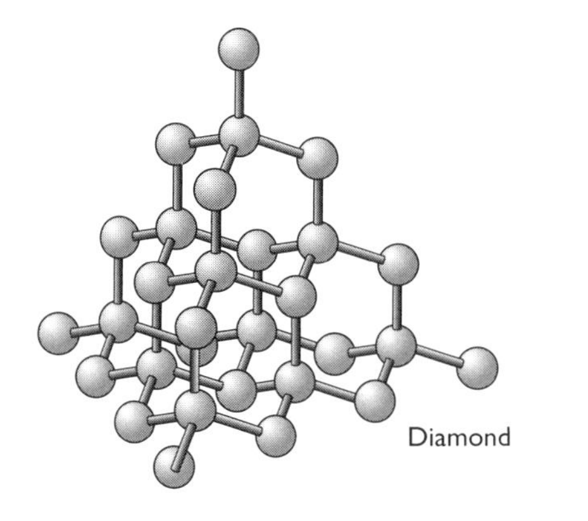

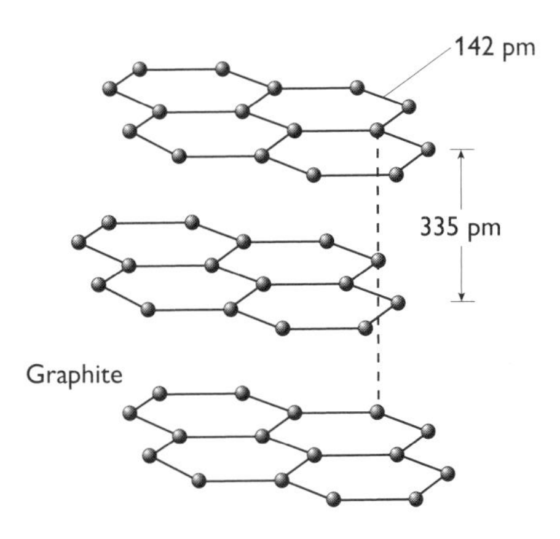

Diamond and graphite have different physical properties, which can be explained in terms of their structures.

Property	**Diamond**	**Graphite**
Structure	Tetrahedral Bond angle 109.5° Symmetrical structure held together by strong covalent bonds throughout the lattice	Hexagonal layers Bond angle 120° Strong covalent bonds within layers Weak van der Waals forces between layers
Electrical conductivity	Poor conductor There are no delocalised electrons All outer-shell electrons are used for covalent bonds	Good conductor Free delocalised electrons able to move throughout the structure
Melting point	High melting point, due to strong covalent bonds throughout the lattice	High melting point, due to strong covalent bonds within layers
Hardness	Hard Tetrahedral shape enables external forces to be spread evenly throughout the lattice other	Soft Bonding within each layer is strong but the weak forces between layers allows the layers to slide over each

Some covalent molecules, such as water and iodine, also form simple molecular lattices. The molecules are held in position in the lattice by intermolecular forces that are comparatively easy to break.

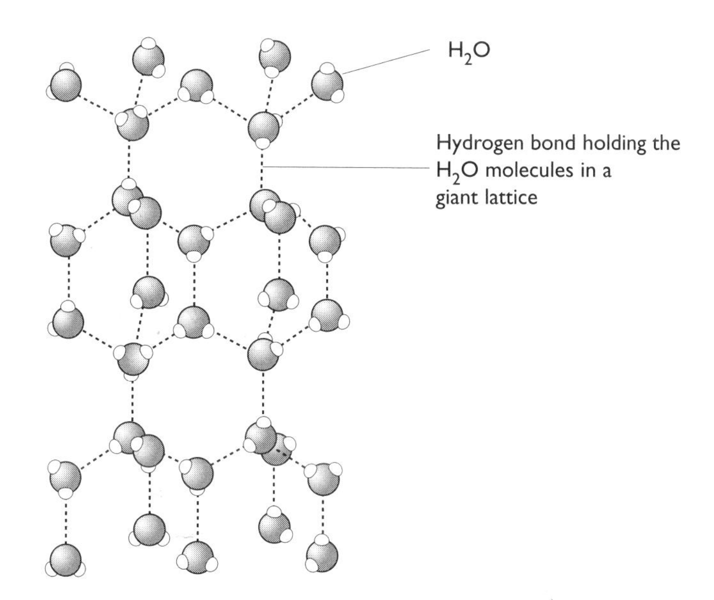

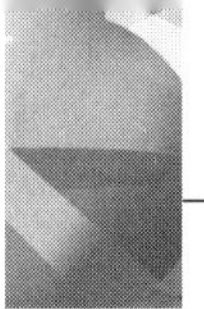

This generally results in low melting points and, in the case of iodine, when the solid is heated it sublimes rather than melts.

Summary

The different types of bonds and forces act as follows:

- covalent bonds act between atoms
- ionic bonds act between ions
- metallic bonds act between positive ions and electrons
- hydrogen bonds act between polar molecules containing hydrogen atoms and either nitrogen, oxygen or fluorine atoms
- permanent dipole–dipole interactions act between polar molecules
- van der Waals forces act between induced dipoles in all molecules and atoms

The periodic table

Periodicity

The periodic table is the arrangement of elements by increasing atomic number. Elements with the same outer-shell electron configuration are grouped together, so physical and chemical properties are periodically repeated.

s block	*d* block	*p* block
	1s	
2s		2*p*
3s		3*p*
4s	3*d*	4*p*
5s	4*d*	5*p*
6s	5*d*	6*p*
7s		

f block
4*f*
5*f*

Trends in the periodic table

Atomic radius

This decreases across a period, because the attraction between the nucleus and outer electrons increases. This is because:

- the nuclear charge increases
- outer electrons are being added to the same shell, so there is no extra shielding

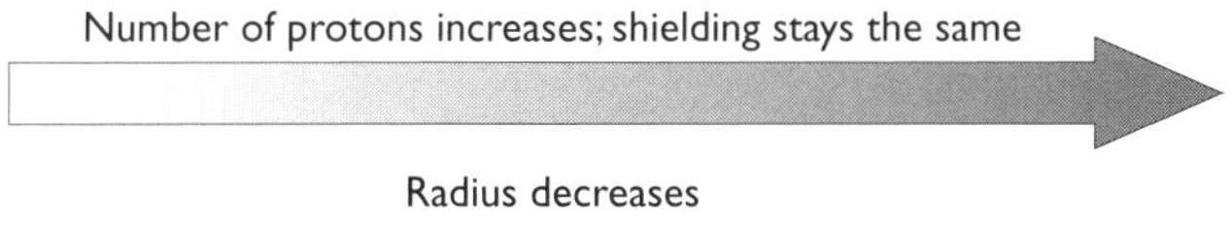

Number of protons increases; shielding stays the same

Radius decreases

Atomic radius increases down a group because the attraction between the nucleus and outer electrons decreases. This is because:

- extra shells are added, resulting in the outer shell being further from the nucleus
- more shells between the outer electrons and the nucleus mean greater shielding

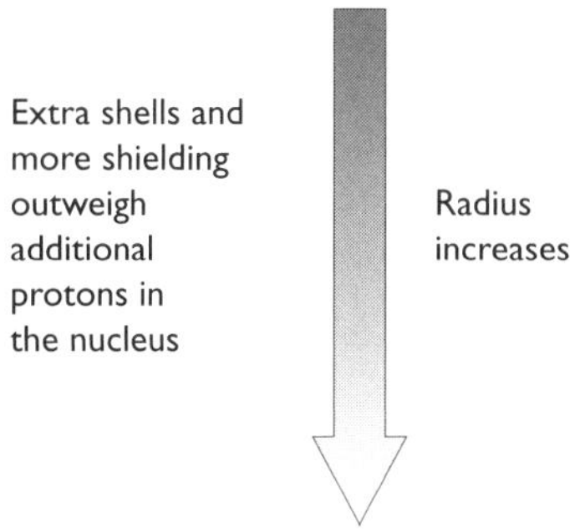

Extra shells and more shielding outweigh additional protons in the nucleus

Radius increases

Electrical conductivity, melting point and boiling point are related to structure and bonding, as shown in the table below.

Giant structures				Molecular structures		
Na	Mg	Al	Si	P_4	S_8	Cl_2
Strong forces between atoms				*Weak* van der Waals forces between molecules		
Metallic			Covalent	van der Waals		
High melting points				Low melting points		
Good conductors			Poor conductors			

Electrical conductivity

Elements in groups 1, 2 and 3 are metals. They are good conductors. Metallic elements are good conductors because they contain mobile, free electrons. The outer-shell electrons can contribute to the mobile, delocalised electrons, allowing metals to conduct heat and electricity, even in the solid state.

The elements in remaining groups across periods 2 and 3 are poor conductors because they do not have any mobile, free electrons. (Graphite is an exception to this and is a good conductor because it has mobile, free electrons.)

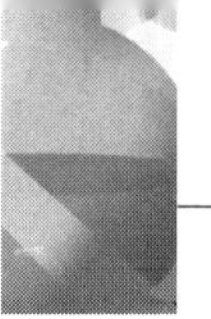

Group 2 elements tend to be better conductors than group 1, because they have two outer-shell electrons, while group 1 elements only have one outer-shell electron.

Melting points and boiling points

These show a gradual increase from group 1 to group 4, followed by a sharp drop to groups 5, 6 and 7. This drop signifies the move from giant structures in groups 1–4, to simple molecular structures in groups 5–7.

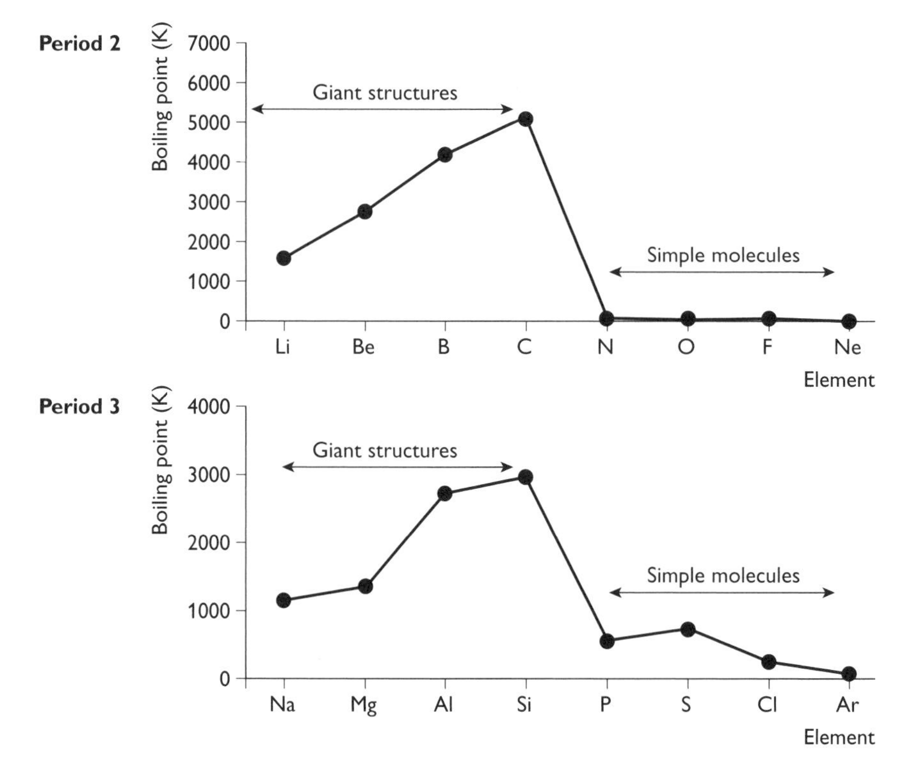

Ionisation energy

Ionisation energy decreases down a group because the outer electrons are further from the nucleus and are shielded by additional inner shells. This results in a decrease in the effective nuclear charge.

Ionisation energy increases across a period due to an increase in nuclear charge and a decrease in atomic radii. Within this general trend there are small 'dips' after group 2 and after group 5.

Group 2

Properties of group 2 elements

Electronic configuration

Each group 2 element has two electrons in its outer shell and readily forms a 2+ ion, which has the same electronic configuration as a noble gas.

$_{12}Mg$ $1s^2\ 2s^2\ 2p^6\ 3s^2$
$_{12}Mg^{2+}$ $1s^2\ 2s^2\ 2p^6$

$_{20}Ca$ $1s^2\ 2s^2\ 2p^6\ 3s^2\ 3p^6\ 4s^2$
$_{20}Ca^{2+}$ $1s^2\ 2s^2\ 2p^6\ 3s^2\ 3p^6$

$_{38}Sr$ $1s^2\ 2s^2\ 2p^6\ 3s^2\ 3p^6\ 3d^{10}\ 4s^2\ 4p^6\ 5s^2$
$_{38}Sr^{2+}$ $1s^2\ 2s^2\ 2p^6\ 3s^2\ 3p^6\ 3d^{10}\ 4s^2\ 4p^6$

$_{56}Ba$ $1s^2\ 2s^2\ 2p^6\ 3s^2\ 3p^6\ 3d^{10}\ 4s^2\ 4p^6\ 4d^{10}\ 5s^2\ 5p^6\ 6s^2$
$_{56}Ba^{2+}$ $1s^2\ 2s^2\ 2p^6\ 3s^2\ 3p^6\ 3d^{10}\ 4s^2\ 4p^6\ 4d^{10}\ 5s^2\ 5p^6$

Physical properties

Group 2 elements are metallic and are therefore good conductors. They have reasonably high melting and boiling points. They generally form colourless ionic compounds that tend to:

- have reasonably high melting and boiling points
- be soluble in water
- be good conductors when molten or aqueous, but poor conductors when solid

Redox reactions of group 2 elements

You should be able to use oxidation numbers to illustrate the redox reactions that occur when group 2 elements react with oxygen and with water. Redox is covered on pages 25–28.

Reactions such as $2Mg(s) + O_2(g) \rightarrow 2MgO(s)$ are known as redox reactions.

Oxidation and reduction can be defined in terms of electrons or in terms of oxidation number.

- **Electrons** — **o**xidation **i**s the **l**oss of electrons; **r**eduction **i**s the **g**ain of electrons (OILRIG).
- **Oxidation number** — oxidation is an increase in oxidation number; reduction is a decrease.

These definitions can both be applied to the reaction below.

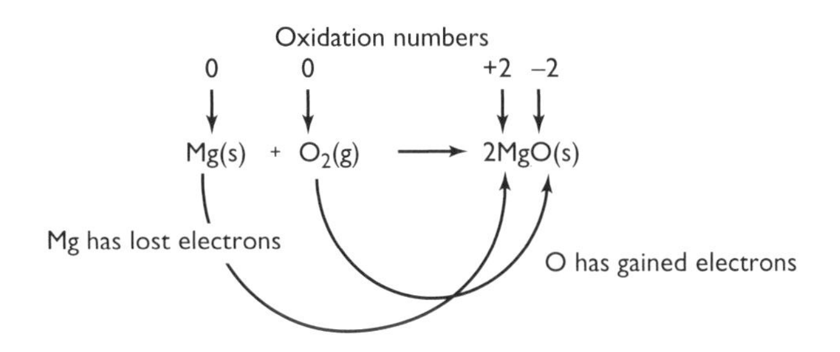

Reaction of group 2 elements with oxygen

Calcium, strontium and barium also react with oxygen to produce their oxides, but reactivity increases down the group. This is explained by the increasing ease with

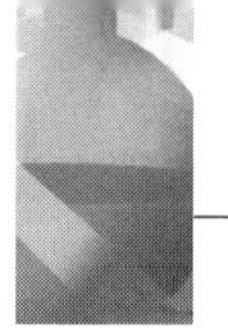

which the group 2 elements form the corresponding 2+ ion.

$Mg(s) + ½O_2(g) \rightarrow MgO(s)$	burns with a bright white light
$Ca(s) + ½O_2(g) \rightarrow CaO(s)$	burns with a brick-red colour
$Sr(s) + ½O_2(g) \rightarrow SrO(s)$	burns with a crimson colour
$Ba(s) + ½O_2(g) \rightarrow BaO(s)$	burns with a green colour

Each of the above reactions is a redox reaction, in that the oxidation number of the group 2 element increases from 0 to +2, while the oxidation number of oxygen decreases from 0 to –2.

Reaction of group 2 elements with water

Group 2 elements also undergo a redox reaction with water:

$Mg(s) + 2H_2O(l) \rightarrow Mg(OH)_2(s) + H_2(g)$

$Ca(s) + 2H_2O(l) \rightarrow Ca(OH)_2(aq) + H_2(g)$

$Sr(s) + 2H_2O(l) \rightarrow Sr(OH)_2(aq) + H_2(g)$

$Ba(s) + 2H_2O(l) \rightarrow Ba(OH)_2(aq) + H_2(g)$

Each of these reactions is also a redox reaction. Once again the oxidation number of the group 2 element increases from 0 to +2, but this time the oxidation number of hydrogen changes from +1 to 0.

Oxidation numbers: Mg 0; in H_2O: H +1, O –2; in $Mg(OH)_2$: Mg +2, O –2, H +1; in H_2: 0

$$\underset{0}{Mg}(s) + 2\overset{+1}{H_2}\overset{-2}{O}(l) \longrightarrow \overset{+2}{Mg}(\overset{-2}{O}\overset{+1}{H})_2(s) + \overset{0}{H_2}(g)$$

The reaction between magnesium and water is very slow, and the resultant $Mg(OH)_2$ is barely soluble in water, forming a white suspension.

Magnesium also reacts with steam, to produce magnesium oxide and hydrogen:

$Mg(s) + H_2O(g) \rightarrow MgO(s) + H_2(g)$

The rate of reaction increases down the group, largely due to the ease of cation (M^{2+}) formation.

Reaction of group 2 oxides with water

All group 2 metal oxides react with water to form hydroxides:

$MgO(s) + H_2O(l) \rightarrow Mg(OH)_2(s)$	magnesium hydroxide is a suspension.
$CaO(s) + H_2O(l) \rightarrow Ca(OH)_2(aq)$	calcium hydroxide solution is limewater
$SrO(s) + H_2O(l) \rightarrow Sr(OH)_2(aq)$	
$BaO(s) + H_2O(l) \rightarrow Ba(OH)_2(aq)$	

These are *not* redox reactions. The oxidation numbers of all the elements remain the same.

Oxidation numbers $\overset{+2\,-2}{MgO(s)} + \overset{+1\,-2}{H_2O(l)} \longrightarrow \overset{+2\,-2\,+1}{Mg(OH)_2(s)}$

The resulting hydroxide solutions are alkaline and have pH values in the region of 8–12. The pH varies according to the concentration of the solution.

Calcium hydroxide is used in agriculture to neutralise acidic soils, while magnesium hydroxide is used in some indigestion tablets as an antacid.

Thermal decomposition of group 2 carbonates

The group 2 carbonates decompose to form oxides and carbon dioxide:

$MgCO_3(s) \rightarrow MgO(s) + CO_2(g)$

$CaCO_3(s) \rightarrow CaO(s) + CO_2(g)$

$SrCO_3(s) \rightarrow SrO(s) + CO_2(g)$

$BaCO_3(s) \rightarrow BaO(s) + CO_2(g)$

The ease with which the carbonates decompose decreases down the group.

Beryllium carbonate is so unstable that it does not exist at room temperature, but barium carbonate requires strong heating to bring about decomposition. Ba^{2+} is a large ion. It has a relatively low charge density, so it does not polarise the CO_3^{2-} ion and therefore does not weaken the ionic bond.

Again, these are *not* redox reactions. The oxidation numbers of all the elements remain the same.

Oxidation numbers $\overset{+2\,+4\,-2}{MgCO_3(s)} \longrightarrow \overset{+2\,-2}{MgO(s)} + \overset{+4\,-2}{CO_2(g)}$

Group 7

Electronic configuration

Each group 7 element has seven electrons in its outer shell and readily forms a 1– ion (an anion) with the same electronic configuration as a noble gas:

$_{9}F$ $1s^2\ 2s^2\ 2p^5$
$_{9}F^-$ $1s^2\ 2s^2\ 2p^6$

$_{17}Cl$ $1s^2\ 2s^2\ 2p^6\ 3s^2\ 3p^5$
$_{17}Cl^-$ $1s^2\ 2s^2\ 2p^6\ 3s^2\ 3p^6$

$_{35}Br$ $1s^2\ 2s^2\ 2p^6\ 3s^2\ 3p^6\ 3d^{10}\ 4s^2\ 4p^5$
$_{35}Br^-$ $1s^2\ 2s^2\ 2p^6\ 3s^2\ 3p^6\ 3d^{10}\ 4s^2\ 4p^6$

$_{53}I$ $1s^2\ 2s^2\ 2p^6\ 3s^2\ 3p^6\ 3d^{10}\ 4s^2\ 4p^6\ 4d^{10}\ 5s^2\ 5p^5$
$_{53}I^-$ $1s^2\ 2s^2\ 2p^6\ 3s^2\ 3p^6\ 3d^{10}\ 4s^2\ 4p^6\ 4d^{10}\ 5s^2\ 5p^6$

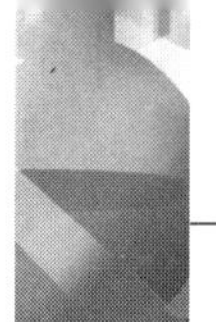

Physical properties

Element	Volatility
Fluorine, F_2 (gas) — yellow	Going down the group, there is an increase in van der Waals forces, corresponding to an increased number of electrons in the halogen molecules. The increase in van der Waals forces reduces the volatility, thus increasing the melting and boiling points
Chlorine, Cl_2 (gas) — green	
Bromine, Br_2 (liquid) — orange/brown	
Iodine, I_2 (solid) — brown/black	
The halogens are all non-metallic, making them poor conductors of electricity	

Chemical properties

The reactivity of the group 7 halogens decreases down the group (opposite to the trend for the group 2 metals, which react by losing electrons). Further down the group it also becomes successively easier to lose electrons (see page 42).

The halogens react not by losing electrons but by gaining an electron to form a halide anion. The ease with which the electron is gained decreases down the group, because atomic radius and shielding both increase down the group, reducing the effective attraction of the nucleus for electrons. Fluorine is the most reactive of the halogens. It is a powerful oxidising agent and readily gains electrons:

$F_2 + 2e^- \rightarrow 2F^-$

$Cl_2 + 2e^- \rightarrow 2Cl^-$

$Br_2 + 2e^- \rightarrow 2Br^-$

$I_2 + 2e^- \rightarrow 2I^-$

Iodine is the least reactive.

It is important to know the difference between a halogen and a halide. In examinations, many students confuse chloride with chlorine. A halogen (fluorine, chlorine and bromine) will displace a halide (Cl^-, Br^- and I^-) from one of its salts. This is shown clearly in the table below:

	Fluoride, F^-	Chloride, Cl^-	Bromide, Br^-	Iodide, I^-
Fluorine, F_2		✓	✓	✓
Chlorine, Cl_2	✗		✓	✓
Bromine, Br_2	✗	✗		✓
Iodine, I_2	✗	✗	✗	

✓ reaction takes place
✗ no reaction

Displacement reactions of the halogens

- Fluorine displaces chloride, bromide and iodide ions from solution.
- Chlorine displaces bromide and iodide ions.
- Bromine only displaces iodide ions.
- Iodine does not displace any of the halides above.

This trend illustrates the decrease in oxidising power down the group.

Chlorine oxidises both bromide and iodide ions:

$Cl_2(aq) + 2Br^-(aq) \rightarrow 2Cl^- + Br_2(l)$

During the above reaction the formation of the orange/brown colour of bromine can be observed.

$Cl_2(aq) + 2I^-(aq) \rightarrow 2Cl^- + I_2(s)$

Bromine oxidises iodide ions only:

$Br_2(aq) + 2I^-(aq) \rightarrow 2Br^- + I_2(s)$

During the above reaction you will see the brown/black colour of iodine. If you add an organic solvent, you will obtain a distinctive violet colour.

Iodine does *not* oxidise either chloride or bromide ions.

The displacement reactions are redox reactions. In each case, the halogen higher in the group gains electrons (is reduced) to form the corresponding halide.

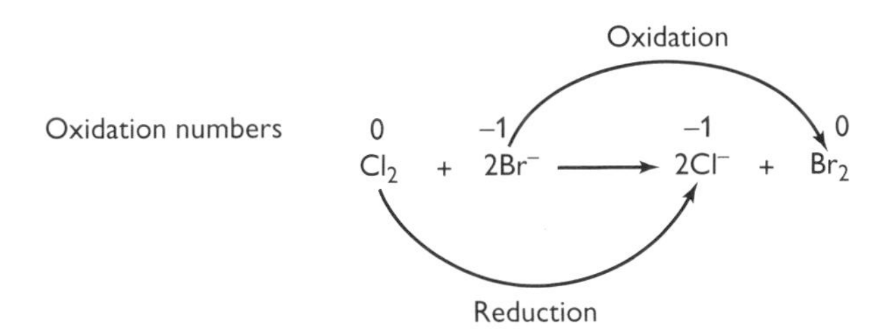

Testing for halide ions

Silver chloride, bromide and iodide are insoluble in water and therefore the chloride, bromide and iodide can be detected by the addition of a solution of silver nitrate ($AgNO_3(aq)$). Each of the silver halides forms a different-coloured precipitate. The precipitates can be distinguished by their solubility in ammonia. AgCl is a white precipitate, which is soluble in dilute ammonia.

$Ag^+(aq) + Cl^-(aq) \rightarrow AgCl(s)$

AgBr is a cream precipitate, which is soluble in concentrated ammonia.

$Ag^+(aq) + Br^-(aq) \rightarrow AgBr(s)$

AgI is a yellow precipitate, which is insoluble in concentrated ammonia.

$Ag^+(aq) + I^-(aq) \rightarrow AgI(s)$

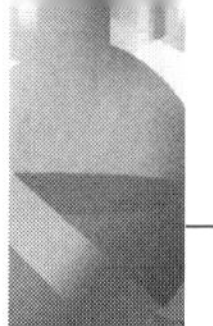

Uses of chlorine

Chlorine is used in water treatment. The gas reacts with water in a reversible reaction, and the resultant mixture kills bacteria, making the water safe to drink. However, chlorine can also react with hydrocarbons in the water, forming chlorinated hydrocarbons, which present a health risk.

$Cl_2 + H_2O \rightleftharpoons HCl + HClO$

The reaction of chlorine with water is a redox reaction, but is unusual in the sense that the chlorine atom undergoes both oxidation and reduction.This is known as **disproportionation**.

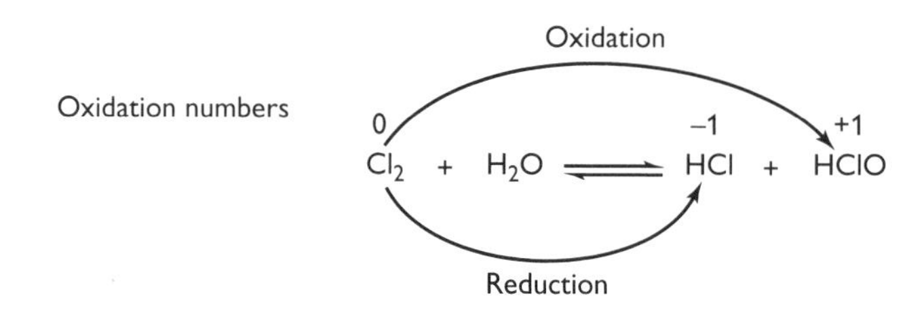

One chlorine atom in the Cl_2 molecule is oxidised as the oxidation number changes from 0 to +1, while the other is reduced from 0 to −1.

Chlorine can also react with NaOH to form bleach. In this reaction the chlorine also undergoes disproportionation.

$Cl_2 + 2NaOH \rightarrow NaCl + NaClO + H_2O$

Questions
&
Answers

This section contains questions similar in style to those you can expect to find in your AS examinations.

The Unit F321 examination lasts 60 minutes and there are 60 marks available. As you can imagine, time is tight, so it is important that you practise answering questions under timed conditions. Each question in this section identifies the specification topic, the total marks and a suggested time that should be spent writing out the answer.

The limited number of questions means that it is impossible to cover all the topics and question styles, but they should give you a flavour of what to expect. The responses that are shown are real students' answers to the questions. Candidate A is an A/B-grade student and candidate B is a B/C-grade student.

There are several ways of using this section. You could:

- 'hide' the answers to each question and try the question yourself. It needn't be a memory test — use your notes to see if you can make all the necessary points
- check your answers against the candidates' responses and make an estimate of the likely standard of your response to each question
- take on the role of the examiner and mark each candidate's response, then check whether you agree with the marks awarded by the examiner
- check your answers against the examiner's comments to see if you can appreciate where you might have lost marks.

Examiner's comments

All candidate responses are followed by examiner's comments, indicated by the icon *e*, which highlight where credit is due. In the weaker answers, they also point out areas for improvement; specific problems; and common errors such as lack of clarity, irrelevance, misinterpretation of the question and mistaken meanings of terms.

Isotopes; electron configuration; ionisation energy

Time allocation: 10–12 minutes

A sample of potassium contains two isotopes: ^{39}K and ^{41}K.

(a) (i) Explain what is meant by the term 'isotope'. (1 mark)

(ii) What is the difference between the two isotopes of potassium? (1 mark)

(iii) The relative atomic mass of the potassium is 39.1. Define the term 'relative atomic mass'. (2 marks)

(b) (i) Write the full electronic configuration of a potassium atom. (1 mark)

(ii) Explain what you understand by the first ionisation energy of potassium. (3 marks)

(iii) Write an equation, including state symbols, for the second ionisation of potassium. (2 marks)

(iv) Explain why the ionisation energies of ^{39}K and ^{41}K are the same. (1 mark)

(c) The first and second ionisation energies of potassium are 419 and 3051 kJ mol^{-1} respectively. Why is there a large difference between the first and the second ionisation energies of potassium? (3 marks)

Total: 14 marks

Candidates' answers to Question 1

Candidate A

(a) (i) Isotopes are atoms of the same element that have the same number of protons but a different number of neutrons.

Candidate B

(a) (i) Same number of protons, different number of neutrons.

Each candidate gains the mark. Candidate A has written a sentence paying due regard to spelling, punctuation and grammar. However, in such questions, marks are not awarded for quality of written communication (QWC), so you can be brief and just stick to the facts.

Candidate A

(a) (ii) ^{41}K has two more neutrons than ^{39}K.

Candidate B

(a) (ii) ^{41}K is heavier than ^{39}K.

Candidate A scores the mark. Candidate B is unlikely to score because he/she has missed out simple quantitative detail. Always try to be as precise as possible. If numbers are given in a question, you usually have to use them in your answer.

Candidate A

(a) (iii) The relative atomic mass is the weighted mean mass of an atom of the element compared to $\frac{1}{12}$ of the mass of an atom of carbon-12, which is taken as exactly 12.

Candidate B

(a) (iii) It is the average mass based on the carbon-12 scale.

Candidate A has learnt the definition of relative atomic mass and gains both marks. The definition given by Candidate B does not take into account the amount of each isotope. If this definition is applied to this sample of potassium, then the average mass of the two isotopes (one with mass 39 and the other 41) is 40, not 39.1. Candidate B scores 1 mark.

Candidate A

(b) (i) $1s^2\ 2s^2\ 2p^6\ 3s^2\ 3p^6\ 4s^1$

Candidate B

(b) (i) 2, 8, 8, 1

Candidate A scores the mark, but Candidate B does not. The full electronic configuration requires subshells as well as principal shells.

Candidate A

(b) (ii) The first ionisation energy is the energy required to remove one electron from one mole of atoms in the gaseous state at s.t.p.

Candidate B

(b) (ii) It is the loss of an electron from the atom in the gaseous state.

Candidate A gains 2 of the 3 marks available; Candidate B scores only 1 mark. Candidate A implies that only a single electron would be removed from 1 mol (6.02×10^{23}) of atoms. Writing an equation can help to put the key points into words. Candidate B understands the concept but has not bothered to learn the detail. This can be costly.

Candidate A

(b) (iii) $K^+(g) \rightarrow K^{2+} + e^-$

Candidate B

(b) (iii) $K(g) \rightarrow K^{2+}(g) + 2e^-$

Each candidate scores only 1 mark. Candidate A has been careless and has lost a mark by forgetting to write the state symbol (g) after K^{2+}. Candidate B earns 1 mark for the state symbols. However, the equation is incorrect as it includes both the first and second ionisations.

Candidate A

(b) (iv) They have the same electron arrangement and same number of protons.

Candidate B

(b) (iv) The only difference is the number of neutrons and they are neutral.

This is a difficult question, which both candidates have answered reasonably well. They each gain the mark, even though neither has fully answered the question. Ionisation energy depends on overcoming the attraction between the electron and the nucleus. This attraction is the same for both isotopes of potassium.

Candidate A

(c) The first electron is easy to remove because it is further from the nucleus and is shielded by an extra inner shell. The second ionisation energy is high because the potassium now has a stable noble gas configuration.

Candidate B

(c) The first ionisation is easy because the electron is further from the nucleus and has more shielding. The second electron is harder to remove because the nucleus is bigger.

Examiners are looking for three key points: distance from the nucleus, shielding and how these affect the attraction between the nucleus and the electrons. The final key point is often either not included or is misinterpreted by stating that 'the nucleus gets bigger' — the nucleus remains the same size and the number of electrons changes. The best way to answer this question is to compare the potassium atom, K, with the potassium ion, K^+. The K^+ ion is smaller than the potassium atom, hence the electrons are held more tightly. The K^+ ion has fewer inner shells/less shielding than the K atom, hence the electrons are held more tightly. Both candidates score 2 marks.

Candidate A scores 11 out of 14 marks and Candidate B scores 7 marks. If Candidate A maintained this standard throughout an examination paper, he/she would be just below the grade-A threshold. Candidate B's mark is equivalent to a grade D. However, with a little care this score could have been pushed up by 3 or 4 marks, which, if maintained throughout a paper, could have a dramatic effect on the final grade.

Moles; test for a halide; decomposition of carbonates

Time allocation: 13–16 minutes

A student prepared magnesium chloride, $MgCl_2$, by adding 8.43 g of magnesium carbonate to 2.00 mol dm^{-3} hydrochloric acid.

$$MgCO_3(s) + 2HCl(aq) \rightarrow MgCl_2(aq) + H_2O(l) + CO_2(g)$$

(a) Suggest two observations that could be made during this reaction. (2 marks)

(b) (i) What amount, in moles, of $MgCO_3$ was used in the experiment? (2 marks)

(ii) Calculate the volume of 2.00 mol dm^{-3} hydrochloric acid needed to react completely with this amount of magnesium carbonate. (2 marks)

(iii) Calculate the volume of CO_2 gas that would be produced at r.t.p. (2 marks)

(c) What reagent(s) would you use to show the presence of chloride ions? State what you would expect to observe. Write an ionic equation for the reaction. (4 marks)

(d) When the carbonates of magnesium, calcium and barium are heated, they decompose and produce an oxide and carbon dioxide.

(i) Write an equation, including state symbols, for the decomposition of one of these carbonates. (2 marks)

(ii) Explain the trend in the ease of decomposition of these carbonates. (3 marks)

Total: 17 marks

Candidates' answers to Question 2

Candidate A

(a) Bubbles of carbon dioxide are given off.

Candidate B

(a) The white solid fizzes and a colourless solution is formed.

e Candidate A scores 1 mark, but Candidate B scores both. The clue is in the question. By looking carefully at the state symbols, you should be able to deduce what you would see. Candidate A has ignored the instructions in the question, which ask for two observations, hence 2 marks for this section.

Candidate A

(b) (i) 0.100 mol

Candidate B

(b) (i) $M_r = 24.3 + 12.0 + 48.0 = 84.3$

8.43

84.3 = 0.100 mol

Each candidate gains 2 marks. However, Candidate B's technique is better than that of Candidate A. It is always advisable to show the working in any calculation. If a mistake is made, some marks can still be awarded for the method. If you make a mistake but show no working, the examiner cannot award any marks.

Candidate A

(b) (ii) 100

Candidate B

(b) (ii) moles of HCl = 0.100

$0.100 = cV$

$\frac{0.100}{2.00} = V = 0.05\ dm^3$

Candidate A has the correct value but scores only 1 mark because there are no units. Candidate B has taken the reacting ratios to be 1:1 rather than 1:2. However, because he/she has shown some working, 1 mark is awarded for the correct use of $n = cV$.

Candidate A

(b) (iii) $2400\ cm^3$

Candidate B

(b) (iii) moles of $CO_2 = 0.100 \times 24 = 2.4\ dm^3$

Each candidate scores 2 marks, but Candidate A runs the risk of losing marks by not showing any working.

Candidate A

(c) Reagent: $AgNO_3(aq)$

Observation: white precipitate

$Ag^+(aq) + Cl^-(aq) \rightarrow AgCl(s)$

Candidate B

(c) Silver nitrate is the reagent. You would see a white solid.

The equation is $Ag^+ + Cl^- \rightarrow AgCl$.

Candidate A scores all 4 marks. Candidate B loses a mark because there is no reference to water as the solvent. An aqueous solution is essential. If he/she had included state symbols ($Ag^+(aq) + Cl^-(aq) \rightarrow AgCl(s)$), Candidate B would have gained both the equation marks.

Candidate A

(d) (i) $MgCO_3(s) \rightarrow MgO(s) + CO_2(g)$

Candidate B

(d) (i) $CaCO_3(s) \rightarrow CaO(s) + CO_2(g)$

Each candidate scores 2 marks.

Candidate A

(d) (ii) $BaCO_3$ is the most stable because the barium ion is large and doesn't polarise the carbonate ion.

Candidate B

(d) (ii) The stability of the carbonates increases down the group.

Candidate B simply states the trend down the group, rather than following the instruction in the question to explain it, and scores only 1 mark. Candidate A's answer is better and scores 2 of the 3 marks available. Candidate A would have scored the third mark if he/she had explained that polarisation leads to a weakening of the ionic bond, making it easier to decompose.

Both candidates have given good answers. However, they need to revise the thermal stability of carbonates. Candidate B should practise mole calculations using balanced equations and the mole ratio given in the balanced equation. Candidate A would do better by showing all the working in calculations and not merely quoting the answer from the calculator. Candidate A scores 14 marks out of 17; Candidate B scores 13 marks.

Oxidation and reduction; bonding and structure

Time allocation: 11–13 minutes

(a) A chemist reacts oxygen separately with magnesium and with sulfur to form MgO and SO_2 respectively. Write an equation for:

(i) the reaction of magnesium and oxygen (1 mark)

(ii) the reaction of sulfur and oxygen (1 mark)

(b) The reactions in (a) are both redox reactions, in which reduction and oxidation take place. Explain, using the changes in oxidation number for sulfur, whether sulfur undergoes oxidation or reduction. (2 marks)

(c) The chemist adds water to MgO and to SO_2, forming two aqueous solutions. Write equations for the reactions that take place and suggest a value for the pH of each solution. (4 marks)

(d) Magnesium oxide is a solid with a melting point of 2852°C; sulfur dioxide has a melting point of –73°C. Explain, in terms of structure and bonding, why there is such a large difference between the melting points of these two oxides. (5 marks)

In this question, 1 mark is available for the quality of written communication.

(1 mark)

Total: 14 marks

Candidates' answers to Question 3

Candidate A

(a) (i) $Mg(s) + \frac{1}{2}O_2(g) \rightarrow MgO(s)$

Candidate B

(a) (i) $2Mg + O_2 \rightarrow 2MgO$

Each candidate scores the mark.

Candidate A

(a) (ii) $S + O_2(g) \rightarrow SO_2(s)$

Candidate B

(a) (ii) $S + O_2 \rightarrow SO_2$

Candidate A could have lost a mark by missing out the state symbol for sulfur, but state symbols are not required here. However, if you do include them, the examiner may penalise you if they are wrong or partly omitted. The best advice is that unless you are asked for state symbols, don't include them in your answer. Both candidates score the mark.

Candidate A

(b) Initial oxidation state of sulfur is 0 and the final oxidation state of sulfur is +4. Therefore, sulfur has undergone oxidation.

Candidate B

(b) The oxidation state of sulfur changes from 0 to 4.

Candidate A scores both marks but Candidate B only scores 1 because he/she does not state that the sulfur had, therefore, been oxidised. Oxidation number has a sign as well as a value. The minus sign should always be included for negative oxidation numbers. If the oxidation number is positive, the + sign should be written, but if there is no sign, the examiner will assume the number to be positive.

Candidate A

(c) $MgO + H_2O \rightarrow Mg(OH)_2$ — pH of the aqueous solution = 11

$SO_2 + H_2O \rightarrow H_2SO_3$ — pH of the aqueous solution = 2

Candidate B

(c) $MgO + H_2O \rightarrow MgOH_2$ — pH = 9

$SO_2 + 2H_2O \rightarrow H_2SO_4 + H_2$ — pH = 5

Candidate A scores all 4 marks. The pH depends on the concentration, so a value between 8 and 13 is acceptable for $Mg(OH)_2$, as is a value between 1 and 6 for H_2SO_3. Candidate B gets 2 marks for the pH predictions but loses both equation marks. The first equation is almost correct, but the formula of magnesium hydroxide must have brackets around the OH. The second equation is incorrect, as the candidate tries to form H_2SO_4 instead of H_2SO_3.

Candidate A

(d) MgO has a high melting point because it has a giant ionic lattice. SO_2 has a low melting point because it is a covalent molecule with weak intermolecular forces.

Candidate B

(d) MgO has strong ionic bonds throughout the lattice which need a lot of energy to break them. SO_2 has covalent bonds that are weak and, therefore, little energy is required to melt it.

The mark scheme is shown below:

- Marking points for magnesium oxide are:
  - ionic lattice ✓
  - strong bonds throughout ✓
  - requires high energy to break these bonds/melt MgO ✓
- Marking points for sulfur dioxide are:
  - covalent ✓
  - polar ✓
  - weak intermolecular/dipole–dipole/van der Waals forces ✓
  - requires low energy to break these bonds/melt SO_2 ✓

Although there are seven marking points, only 1 mark is available for the amount of energy required. For quality of written communication, the examiner is looking for the correct use of chemical terms such as lattice, intermolecular forces, van der Waals, polar etc. To gain the mark, at least two terms must be used and spelt correctly.

Candidate A scores 3 chemistry marks. The ionic nature of magnesium oxide is awarded 1 mark, but the second mark is lost because there is no mention of the strength of the ionic bonds. Two of the 3 marks available for sulfur dioxide are awarded, but a mark is lost because there is no reference to its polar nature. The amount of energy required to overcome the intermolecular forces is not mentioned. Candidate A makes correct use of the terms 'lattice' and 'intermolecular bonds' and, therefore, scores 1 mark for quality of written communication.

Candidate B scores 4 chemistry marks. All 3 marks are awarded for magnesium oxide, but only 1 mark for the description of sulfur dioxide. The most common incorrect answer to this type of question is demonstrated by Candidate B, who describes covalent bonds as weak. This is not the case — covalent bonds are strong bonds. Candidate B did not gain a mark for quality of written communication because only one term, 'lattice', is used correctly.

The free-response question proves difficult for many students, and here both candidates lose marks. When attempting to answer free-response questions it is essential that you devise a plan based on the information given in the question. There are 6 marks available, one of which is for quality of written communication. This leaves 5 marks for commenting on the melting points of the two substances. It is safe to assume that the examiner will split these marks between the two compounds and it would be sensible to give three points for each as follows:

Magnesium oxide has a giant ionic lattice with strong bonds throughout. Hence, a great deal of energy is needed to melt it.

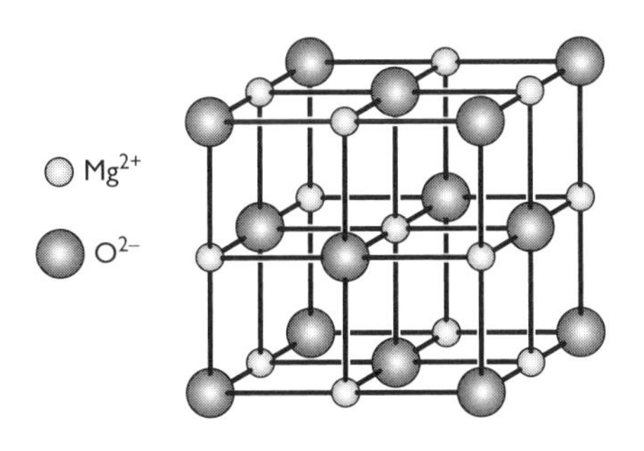

Sulfur dioxide has polar covalent molecules with weak intermolecular forces. Hence, only a small amount of energy is needed to melt it.

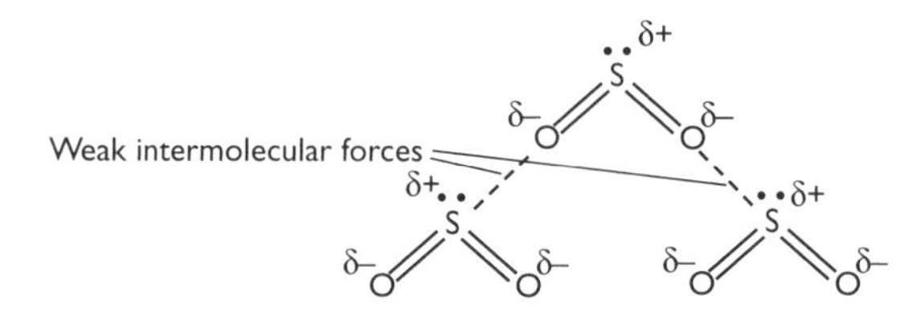

Both candidates have answered the structured parts well, with Candidate A scoring the maximum 8 marks and Candidate B scoring 6. On the free-response question, Candidate A scores 4 marks out of 6 and Candidate B also scores 4.

Overall, Candidate A scores 12 marks and Candidate B scores 10.

Electronegativity; hydrogen bonding

Time allocation: 9–10 minutes

(a) Explain what is meant by the term 'electronegativity'. (2 marks)

(b) (i) Draw a diagram to show hydrogen bonding between two molecules of water. Your diagram must include dipoles and lone pairs of electrons. (4 marks)

(ii) State the bond angle in a water molecule. (1 mark)

(c) State and explain two properties of ice that are a direct result of hydrogen bonding. (4 marks)

Total: 11 marks

Candidates' answers to Question 4

Candidate A

(a) Attraction for electrons within a covalent bond.

Candidate B

(a) Different atoms have different attractions for electrons.

Candidate A gains both marks. Candidate B scores only 1 mark, for a partly correct answer that does not mention the covalent bond.

Candidate A

(b) (i)

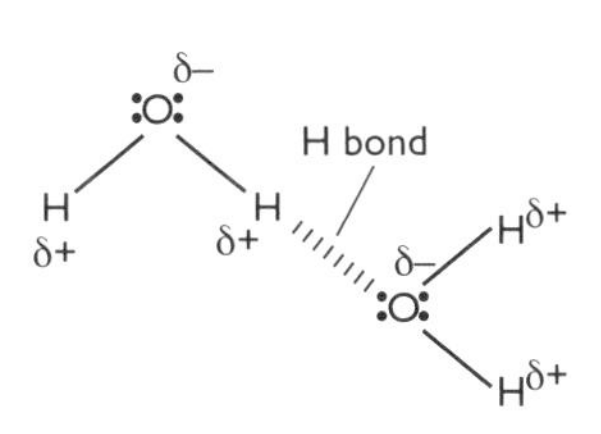

Candidate B

(b) (i)

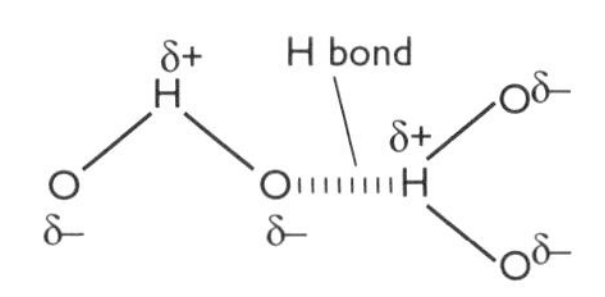

The four marking points are:

- correct dipoles with $O^{\delta-}$ and $H^{\delta+}$ ✓
- water drawn as non-linear ✓
- the hydrogen bond between a hydrogen atom in one water molecule and the oxygen atom in an adjacent water molecule ✓
- the involvement of the lone pair of electrons on oxygen and the linear shape of O—H⋯O ✓

Candidate A gives the perfect answer and earns all 4 marks. To obtain full marks, the hydrogen bond must be drawn carefully so that the sketch shows clearly the involvement of a lone pair of electrons on the oxygen atom, which many candidates fail to do.

A substantial minority of candidates draw water as HO_2, not H_2O. Should Candidate B lose all 4 marks? The dipoles are correct, the shape is correct and a hydrogen bond has been drawn between the hydrogen in one water molecule and the oxygen in another water molecule. However, there is no indication of the involvement of the lone pair of electrons, and the H—O ııııı H is not drawn as linear. Candidate B scores 1 mark (maximum 2), but he/she could have gained 3 marks by drawing the structure of water correctly. It is easy to lose marks through carelessness.

Candidate A

(b) (ii) Approximately 104°

Candidate B

(b) (ii) 104.5°

Both candidates score the mark for this simple recall question.

Candidate A

(c) Ice floats on water because air is trapped between the water molecules. Ice has a higher melting point than expected because of the hydrogen bonds.

Candidate B

(c) Ice is less dense than water because the hydrogen bonds hold the molecules further apart. Ice has a high melting point because of the hydrogen bonds.

Each candidate scores 3 marks, for different reasons. Candidate A gets 2 marks for the two properties of ice, but the explanation for the fact that ice floats on water is incorrect. Candidate B almost gains 4 marks. However, the statement that 'ice has a high melting point' should be qualified by writing 'a higher melting point than expected' in order to earn the mark.

Candidates A and B seem to have similar ability, but often the outcome does not reflect this with Candidate A scoring more marks than Candidate B. This is usually down to either examination technique or carelessness. In this question, Candidate A scores 10 marks out of 11, whereas Candidate B scores only 6 marks. Candidate A could achieve a grade A; Candidate B's scores fluctuate between grade A and grade D. It is useful to look at Candidate B's responses and identify the sort of errors/slips. If you can recognise the mistakes made by others, you may avoid making them yourself.

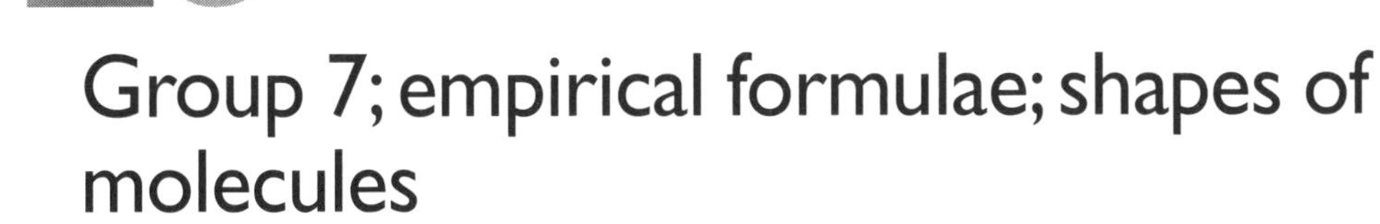

Group 7; empirical formulae; shapes of molecules

Time allocation: 9–10 minutes

(a) Chlorine bleach is made by the reaction of chlorine with aqueous sodium hydroxide. In this reaction the oxidation number of chlorine changes and it is said to undergo disproportionation.

$$Cl_2(g) + 2NaOH(aq) \rightarrow NaClO(aq) + NaCl(aq) + H_2O(l)$$

(i) Determine the oxidation number of chlorine in Cl_2, NaClO and NaCl. (3 marks)

(ii) State what is meant by the term 'disproportionation'. (1 mark)

(iii) The bleaching agent is the ClO^- ion. In the presence of sunlight, this ion decomposes to release oxygen gas. Construct an equation for this reaction. (2 marks)

(b) The sea contains a low concentration of bromide ions. Bromine can be extracted from seawater by first concentrating the seawater and then bubbling chlorine through this solution.

(i) Suggest how seawater could be concentrated. (1 mark)

(ii) The chlorine oxidises bromide ions to bromine. Construct a balanced ionic equation for this reaction. (1 mark)

(c) Vinyl chloride is a compound of chlorine, carbon and hydrogen. It is used to make polyvinylchloride (PVC). Vinyl chloride has the following percentage composition by mass: chlorine 56.8%; carbon 38.4%; hydrogen 4.8%.

(i) Show that the empirical formula of vinyl chloride is C_2H_3Cl. Show your working. (2 marks)

(ii) The molecular formula of vinyl chloride is the same as its empirical formula.

Draw a possible structure, including bond angles, for a molecule of vinyl chloride. (2 marks)

Total: 12 marks

Candidates' answers to Question 5

Candidate A

(a) (i) $Cl_2(g) + 2NaOH(aq) \rightarrow NaClO(aq) + NaCl(aq) + H_2O(l)$

0 +1 −1

Candidate B

(a) (i) $Cl_2(g) + 2NaOH(aq) \rightarrow NaClO(aq) + NaCl(aq) + H_2O(l)$

0 1 1

Candidate A scores all 3 marks and Candidate B scores 2 marks. Oxidation number has a sign as well as a value. It is always necessary to include the minus sign for negative oxidation numbers. If the oxidation number is positive, the '+' sign should be written, but the examiner will assume that the number is positive if no sign is given.

Candidate A

(a) (ii) Chlorine is not proportional.

Candidate B

(a) (ii) Chlorine will displace bromine and iodine.

Neither candidate scores the mark. The question command word 'State...' indicates that this is recall and that you should have learnt the definition. In Unit F321, disproportionation is defined as 'a reaction in which an element is simultaneously oxidised and reduced'.

Candidate A

(a) (iii) $ClO^- \rightarrow Cl^- + \frac{1}{2}O_2$

Candidate B

(a) (iii) $2ClO^- \rightarrow 2Cl + O_2$

Candidate A scores both marks, while Candidate B loses a mark by not balancing the equation for charge. There is a net charge of 2– on the left-hand side of the equation, so there must also be a net charge of 2– on the right-hand side. The 2Cl should be $2Cl^-$.

Candidate A

(b) (i) Evaporation

Candidate B

(b) (i) Heat

Both candidates score the mark because both methods would work.

Candidate A

(b) (ii) $Cl_2 + 2Br^- \rightarrow Br_2 + 2Cl$

Candidate B

(b) (ii) $Cl_2 + Br_2 \rightarrow 2Cl^- + 2Br^-$

Neither candidate scores the mark. Candidate A has made the same mistake that Candidate B made in part (a)(iii) and has forgotten to balance the net charges. Candidate B has completely confused bromine and bromide, chlorine and chloride. The instructions in the question tell you exactly what is happening but you have to read the question carefully.

Questions on displacement reactions should be straightforward recall of knowledge. However, candidates seem to find ionic equations difficult and even able candidates get them wrong. Many students are confused by the names 'bromide' and 'bromine'. It is worth spending some time ensuring that you know and understand these reactions and that you recognise that a word ending in '-ide' is an ion and has a negative charge and a word ending in '-ine' is an atom or a diatomic molecule.

Candidate A

(c) (i) Divide the percentage of each element by its relative atomic mass:

chlorine (56.8/35.5) = 1.6, carbon (38.4/12.0) = 3.2, hydrogen = (4.8/1.0) = 4.8

Ratio is 1.6:3.2:4.8

Divide by the smallest: 1.6/1.6:3.2/1.6:4.8/1.6

Therefore, the ratio is 1:2:3, which is equivalent to ClC_2H_3.

So the empirical formula is C_2H_3Cl.

Candidate B

(c) (i) M_r of ClC_2H_3 is 35.5 + 24.0 + 3.0 = 62.5

amount of Cl = $\frac{56.8 \times 62.5}{100}$ = 35.5 therefore 1 Cl

amount of C = $\frac{38.4 \times 62.5}{100}$ = 24 therefore 2 C

amount of H = $\frac{4.8 \times 62.5}{100}$ = 3 therefore 3 H

Therefore, the ratio is 1:2:3, which is equivalent to ClC_2H_3. So the empirical formula is C_2H_3Cl.

Each candidate scores 2 marks. Candidates A and B use different methods to calculate the empirical formula, but both are valid and hence earn the marks.

Candidate A

(c) (ii)

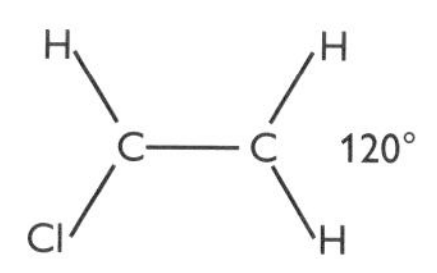

Candidate B

(c) (ii)

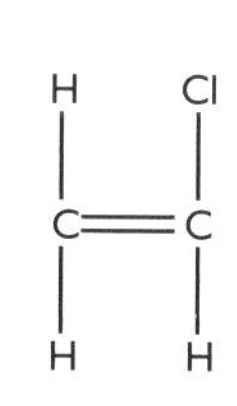

A correct C=C double bond automatically earns 1 mark. The second mark is for the rest of the sketch and the bond angle. Each candidate scores 1 mark. Candidate A loses a mark because there is no C=C double bond; Candidate B loses a mark because the shape is not correct and a bond angle has not been included.

Overall, Candidate A scores 9 out of 12 marks; Candidate B scores 7.

Ionisation energies

Time allocation: 9–10 minutes

Successive ionisation energies can provide evidence for the electronic configuration of an element. The table below provides data on the successive ionisation energies of oxygen.

Ionisation number	1st	2nd	3rd	4th	5th	6th	7th	8th
Ionisation energy (kJ mol^{-1})	1314	3388	5301	7469	10 989	13 327	71 337	84 080
log (ionisation energy)	3.1	3.5	3.7	3.9	4.0	4.1	4.8	4.9

(a) (i) Plot log (ionisation energy) against ionisation number. Explain how these data provide evidence for the electronic configuration of oxygen. (3 marks)

(ii) Explain why the successive ionisation energies of oxygen increase. (2 marks)

(iii) Write an equation to represent the third ionisation energy of oxygen. (2 marks)

(b) The first five ionisation energies of element X are shown below. State, with a reason, the group in which you would expect to find element X. (2 marks)

Ionisation energy	1st	2nd	3rd	4th	5th
kJ mol^{-1}	578	1817	2745	11 578	14 831

(c) Use the grid below to sketch the plot of log (ionisation energy) against ionisation number for phosphorus, $_{15}P$. (3 marks)

Total: 12 marks

Candidates' answers to Question 6

Candidate A

(a) (i)

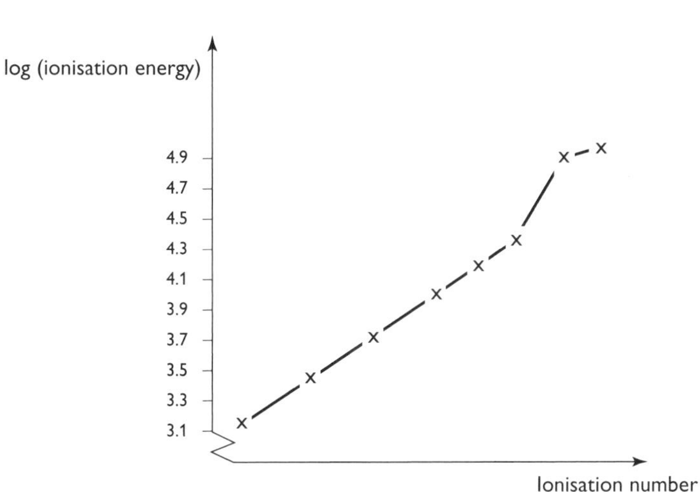

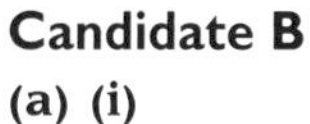

There is a large jump after the sixth ionisation energy. This shows that a new inner shell is present.

Candidate B

(a) (i)

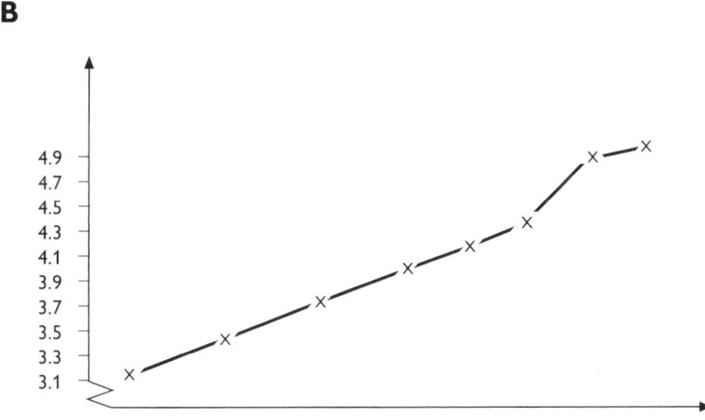

There is a gradual increase in ionisation from one to six. The large increase from six to seven shows a new shell. Hence oxygen has six electrons in its outer shell.

Candidate A earns 2 marks for the graph, but candidate B loses a mark for not labelling the axes. Both are awarded a mark for relating the large increase in ionisation energy, which occurs between the sixth and the seventh ionisation, to evidence for a new shell.

Candidate A

(a) (ii) When electrons are removed, the ionic radii decrease due to an increase in the size of the nucleus.

Candidate B

(a) (ii) Each time the nucleus gets bigger, it is harder to remove the next electron.

Both candidates have misunderstood the process. The nucleus remains the same throughout. Oxygen always has eight protons, and this never changes. What does change is the proton to electron ratio. Each time an electron is removed, the proton to electron ratio changes in favour of the protons. Candidate A picks up 1 mark out of the 2 marks awarded, by pointing out that the radius decreases, but candidate B scores zero. The most important factor is the change in radius — the loss of electrons results in an ion with a smaller radius, making the next electron more difficult to remove because it is located closer to the nucleus. Both candidates seem to understand what is happening, but both lose marks because they do not express themselves precisely. The examiner cannot *interpret* what is written.

Candidate A

(a) (iii) $O^{2+}(g) \rightarrow O^{3+}(g) + e^-$

Candidate B

(a) (iii) $O_2(g) \rightarrow O_2^{3+}(g) + 3e^-$

Candidate A is awarded both marks, but candidate B needs to carefully revise the basic definitions, having made two fundamental errors:

- ionisation energy relates to *atoms* not molecules
- ionisation energy relates to the removal of *one electron at a time* (not three)

Ionisation energy is defined as the removal of one electron from each atom/ion in one mole of gaseous atoms/ions. The general equation is $X^{(n-1)+}(g) \rightarrow X^{n+}(g) + e^-$, where n = 1, 2, 3 etc.

Candidate A

(b) Group 5, because the biggest increase in ionisation energy is between the fifth and sixth ionisation energies. This shows that the sixth electron must be in an inner shell much closer to the nucleus.

Candidate B

(b) 5

Candidate A gets both marks and gives a good, reasoned answer. Candidate B also has the correct answer, but only scores 1 of the 2 marks available. Candidate B needs to read the question carefully and to look at how the marks are allocated.

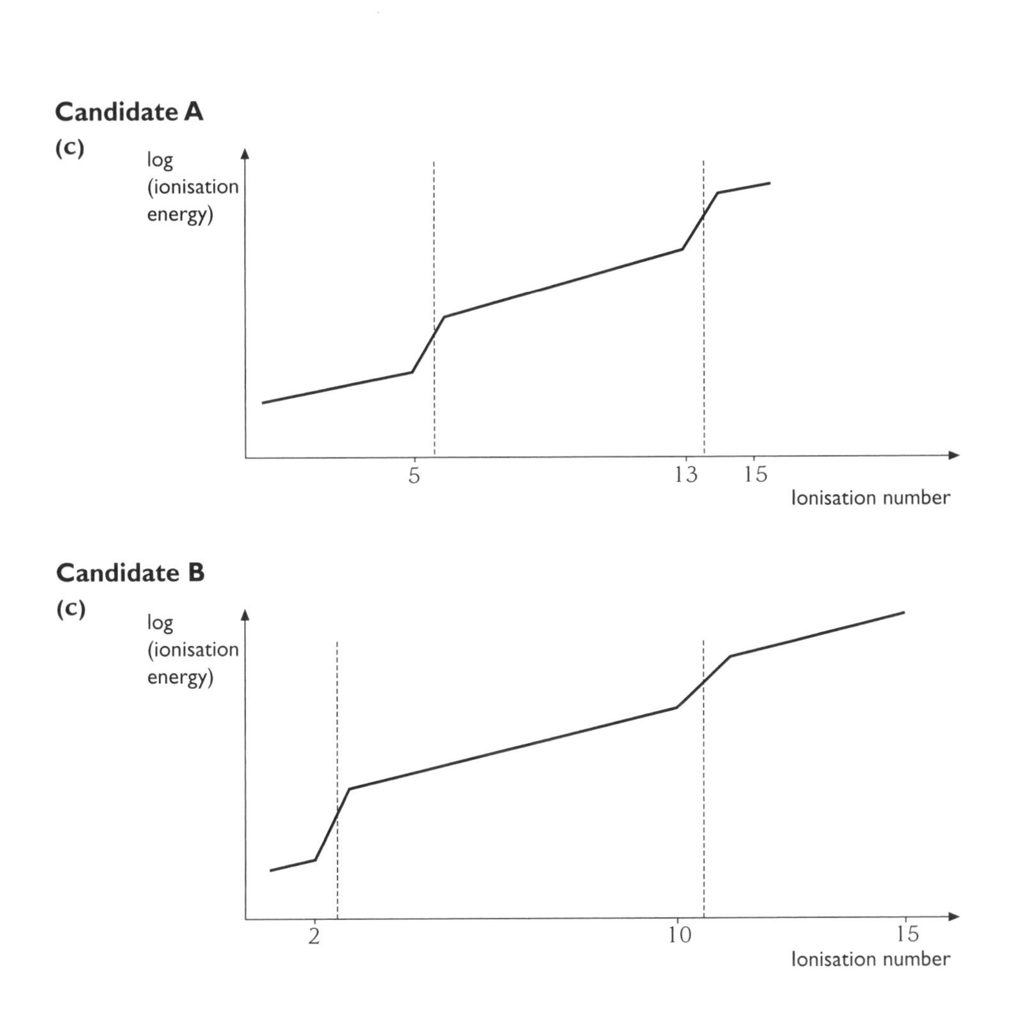

e Marks are awarded for recognising that because phosphorus is in group 5, there will be large increases in the ionisation energy after the fifth and the thirteenth (5 + 8) electrons are removed. Candidate A gains all 3 marks, but candidate B loses 2 marks by drawing the ionisation energies back-to-front. This is a mistake made by many candidates, who tend to think along the lines of $_{15}P$, and therefore use the electronic configuration 2,8,5 and draw a sketch showing that the inner electrons are removed first. It is worthwhile reading through candidate B's answer again, listing ways in which additional marks could have been gained with a little more care.

e **Looking at each answer in isolation, there seems to be little difference in standard between the two candidates. Candidate A is more careful and appears to have learnt the basic principles. Candidate B needs to improve his/her basic knowledge. Overall, candidate A scores 11 marks out of 12 (grade A) and candidate B scores just 4, which is grade E/U.**

Bonding

Time allocation: 12–14 minutes

Chlorine reacts with sodium to form sodium chloride.

(a) Describe the bonding in sodium, chlorine and sodium chloride. (7 marks)

(b) Relate the physical properties of chlorine and sodium chloride to their structures and bonding. (8 marks)

In this question, 1 mark is available for the quality of written communication. (1 mark)

Total: 16 marks

Candidates' answers to Question 7

Candidate A

(a) Sodium contains a giant metallic bond. This is formed by the close packing of the atoms so that their outer-shell electrons overlap and the electrons are free to move anywhere in the lattice. The free electrons are like a 'glue' that holds the lattice together.

Chlorine is a covalent molecule made by sharing electrons.

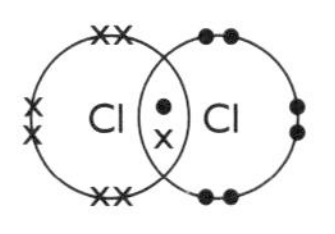

Sodium chloride forms ionic bonds. An ionic bond is the electrostatic attraction between oppositely charged ions which results from the transfer of an electron from the sodium to the chlorine.

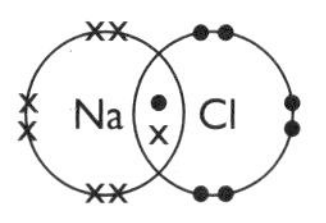

Candidate B

(a)

Sodium	Chlorine	Sodium chloride
Metallic bonding	Covalent bonding	Ionic bonding
e ⊕ e ⊕ e ⊕ e ⊕ e ⊕ e ⊕ e ⊕ e ⊕ e ⊕ e e e	Cl Cl	Na + Cl −
There is a giant lattice of positive ions surrounded by a sea of free electrons	There is a shared pair of electrons in the bond	The sodium transfers an electron to the chlorine and the oppositely charged ions are attracted to each other

The mark scheme is shown below:

- sodium — metallic ✓; lattice of positive ions ✓; delocalised/free electrons ✓
- chlorine — covalent ✓; sharing electrons ✓
- sodium chloride — ionic ✓; electrostatic attraction between oppositely charged ions ✓

Candidate A scores 5 of the 7 marks available. In the bonding of sodium, Candidate A has missed out 'positive ions' in the description of the lattice and the diagram drawn for the structure of sodium chloride contradicts the description in words. Candidate A explains ionic bonding correctly but draws a diagram showing shared electrons, making it covalent. Candidate A's answer is ambiguous — it does not make it clear whether or not NaCl is ionic or covalent. The examiner will not select the correct response for you. If you give two pieces of information that are contradictory, the examiner will always mark the incorrect answer first. There was no need to draw the diagram — if diagrams are required, the question will usually ask for them. Candidate B scores all 7 marks. In these circumstances, a table is an acceptable way of answering the question.

Candidate A

(b) Chlorine is a gas because the bonds between the Cl_2 molecules are very weak van der Waals forces. It is a poor conductor because it contains no mobile electrons.

Sodium chloride has a high melting point because there are strong ionic bonds throughout the lattice. It conducts electricity when molten or aqueous but not when solid. This is because when molten or aqueous the electrons are free to move.

Candidate B

(b)

Substance	Property
Chlorine	Gas; poor conductor; insoluble in water
Sodium chloride	Solid; conducts when molten or aqueous; dissolves in water

The mark scheme is shown below:

- chlorine — two properties (2 marks); two reasons explaining the properties (2 marks)
- sodium chloride — two properties (2 marks); two reasons explaining the properties (2 marks)

Students often find free-response questions difficult. However, there are always clues in the question. Here, you are given two substances, chlorine and sodium chloride, and asked to relate their physical properties to their structures and bonding. There are 8 marks available, so it is logical to give two properties for each along with explanations of each property, i.e. eight different points for 8 marks.

Candidate A scores all 4 marks for chlorine but loses a mark for sodium chloride because sodium chloride conducts electricity by the movement of ions not electrons.

Candidate B has again chosen to use a table, which is a valid way to answer the question. However, he/she has made several errors. Three general properties of covalent substances are listed but the third, with respect to chlorine, is incorrect. Chlorine is slightly soluble in water and reacts with water to produce a mixture of HCl and HClO. Therefore, Candidate B obtains only 1 mark for the properties, even though two properties are correct (remember the third property is wrong). For sodium chloride, Candidate B lists three properties that are all correct and so gains 2 marks. He/she has forgotten to give explanations for the properties and automatically loses 4 marks.

Examiners accept that chemists communicate in various ways: equations, diagrams and tables. However, whenever there are marks for quality of written communication, you must write at least two consecutive sentences that are relevant to the question. Candidate A gains the mark for quality of written communication and scores a total of 13 marks out of 16. Candidate B loses the mark for quality of written communication because the answer contains no continuous prose. Candidate B scores 10 marks.

Trends in atomic radii

Time allocation: 6–7 minutes

The atomic radii of elements in periods 2 and 3 are shown in the table below.

		Group						
		1	**2**	**3**	**4**	**5**	**6**	**7**
Period 2	**Element**	Li	Be	B	C	N	O	F
	Atomic radius/nm	0.134	0.125	0.090	0.077	0.075	0.073	0.071
Period 3	**Element**	Na	Mg	Al	Si	P	S	Cl
	Atomic radius/nm	0.154	0.145	0.130	0.118	0.110	1.102	0.099

(a) (i) State the trend shown in atomic radius across a period. (1 mark)

(ii) Explain this trend. (3 marks)

(b) (i) State the trend shown in atomic radius down a group. (1 mark)

(ii) Explain this trend. (3 marks)

Total: 8 marks

Candidates' answers to Question 8

Candidate A

(a) (i) Atomic radii decrease across the period.

Candidate B

(a) (i) It gets smaller.

Both candidates gain the mark, since for these questions you do not need to write in whole sentences: 'decrease' would be sufficient. Time is very tight, with only 60 minutes allocated for a 60-mark paper, so it is a good idea to write only the minimum amount required.

Candidate A

(a) (ii) The number of protons in the nucleus increases, but the electrons fill up the same shell with the same shielding, and therefore the effective nuclear attraction increases, pulling in the outer electrons.

Candidate B

(a) (ii) The pull of the nucleus increases and the electrons fill up the same main shell.

Candidate A gives a very good answer and scores all 3 marks. Candidate B clearly understands this topic, but has not used the information in the question. Three separate points are required, with each point gaining 1 mark. They are as follows: increased effective nuclear charge; electrons fill up same shell; all elements across the period experience the same degree of shielding from the inner shells.

Candidate A

(b) (i) Increase

Candidate B

(b) (i) Gets bigger

e Both answers score the mark.

Candidate A

(b) (ii) The number of protons in the nucleus increases, but this is compensated by the extra shell and the increased shielding by inner shells. This results in a decrease in the effective nuclear charge.

Candidate B

(b) (ii) There is an extra shell and more shielding, so the outer electrons are not held as tightly.

e Candidate A again provides a good answer and is awarded all 3 marks. Candidate B scores 2 marks. The essential part missing from candidate B's answer is the idea that, despite the increased number of protons in the nucleus, the radius still increases due to the additional shielding.

e **Both candidates clearly understand the concept of variation in atomic radii, but candidate B has not stated or explained fully the reasons for the variation. Candidate A has worked hard and has fully learnt and understood the textbook explanations, and so scores a maximum 8 out of 8 marks. Candidate B scores a very creditable 6 out of 8 marks, but throws away 2 marks by not taking time to reflect on what has been written in the question. Learning how to assess questions will only come with practice — it is a good idea to try timed questions as part of your revision. Don't forget that every mark counts. If candidate B scored 6 out of every 8 marks throughout the exam paper, then this would translate to 45 marks out of 60, which is equivalent to a grade B.**

However, scoring 7 out of 8 marks would correspond to 52 marks out of 60, which is a grade A.

Group 2

Time allocation: 9–10 minutes

(a) Barium is a group 2 element. It reacts with oxygen to form compound A, and with water to form compound B and gas C. With the aid of suitable equations, identify A, B and C. (5 marks)

(b) (i) Write an equation, including state symbols, for the thermal decomposition of barium carbonate. (2 marks)

(ii) Calculate the minimum volume of 0.05 mol dm^{-3} HCl(aq) that would be needed to react with 1.00 g of barium carbonate. Show all your working. (4 marks)

Total: 11 marks

Candidates' answers to Question 9

Candidate A

(a) $Ba + \frac{1}{2}O_2 \rightarrow BaO$; compound A is barium oxide.

$Ba + 2H_2O \rightarrow Ba(OH)_2 + H_2$; compound B is barium hydroxide, and gas C is hydrogen.

Candidate B

(a) $2Ba + O_2 \rightarrow 2BaO$; compound A is BaO.

$Ba + 2H_2O \rightarrow Ba(OH)_2 + H_2$; compound B is $Ba(OH)_2$, and gas C is H_2.

e These are excellent answers from both candidates, and they will score maximum marks. When the question asks you to identify a substance, it can be identified either by name or by formula, as long as it is unambiguous. It is safer to name the compounds, as mistakes are often made when quoting formulae. Equations have to be balanced, and it is acceptable to balance them by using fractions if appropriate.

Candidate A

(b) (i) $BaCO_3(s) \rightarrow BaO(s) + CO_2(g)$

Candidate B

(b) (i) $BaCO_3 \rightarrow BaO + CO_2$

e Both candidates know their chemistry, but candidate A displays better examination technique by carefully reading the question and then following the instructions. Candidate B fails to respond to the instruction to include state symbols, and carelessly loses 1 mark.

Candidate A

(b) (ii) Molar mass of $BaCO_3$ = 137 + 12 + 48 = 197 g mol^{-1}

Moles of $BaCO_3$ = 1/197 = 0.0050761 = 0.005 mol

Moles of HCl needed = 0.005 mol

Volume of HCl = n/c = 0.005/0.05 = 0.1 dm^3 = 100 cm^3

Candidate B

(b) (ii) 203 cm^3

Candidate B gets all 4 marks, but displays an appalling examination technique. If the numerical value had been incorrect, candidate B would have lost all 4 marks. Candidate A's approach is much more sensible. Each step has been clearly shown and can be followed by the examiner. Candidate A has not used the mole ratio when $BaCO_3$ reacts with HCl (the ratio is 1:2). This one error in the calculation would therefore only lose 1 mark. Candidate A's technique is not good because he/she has rounded up in the middle of the calculation, and it is best to keep the numbers in the calculator until the final step and then round to an appropriate number of significant figures. The correct calculator value is 101.522842, which to 3 significant figures is 102 cm^3. This would cost candidate A a second mark. Candidate A scores 2 marks.

Overall, candidate A scores 9 marks out of 11 and candidate B scores 10.

Shapes of molecules

Time allocation: 8–10 minutes

Electron-pair repulsion theory can be used to predict the shape of covalent molecules. State what is meant by the term 'electron-pair repulsion theory' and use it to determine the shapes of four molecules of your choice. Choose molecules to illustrate four different shapes. State the bond angle in each shape. (10 marks)

In this question, 1 mark is available for the quality of written communication. (1 mark)

Total: 11 marks

Candidates' answers to Question 10

Candidate A

The shape of a molecule depends on the number of electron pairs around the central atom. Each pair repels the others but lone pairs of electrons repel more than bonded pairs.

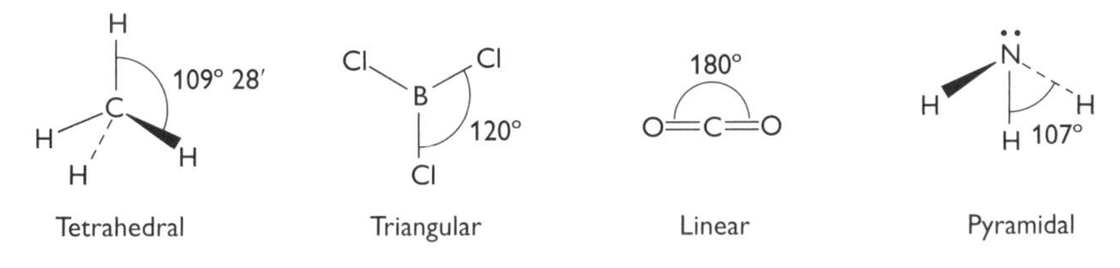

Candidate B

Electron-pair repulsion theory is a theory that tells us about electron-pair repulsion. It tells us that electrons repel each other and therefore electrons do not pair up because they repel each other.

The shape of molecules depends on the number of bonds in a molecule.

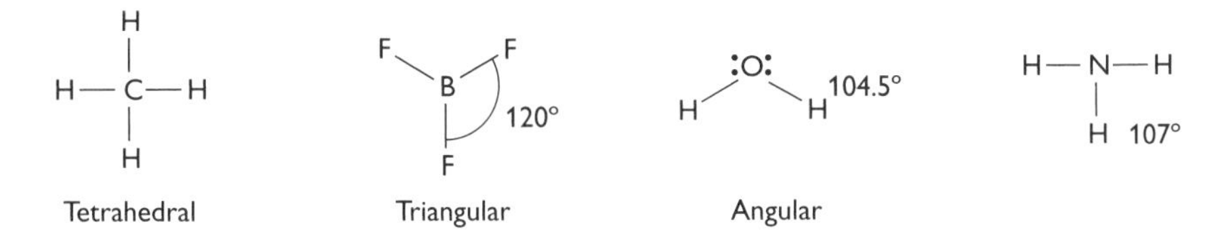

Candidate A gives an excellent answer scoring 2 marks for the description of electron-pair repulsion theory and 2 marks each for the four shapes. The quality of written communication mark is awarded because specific chemical terms are used in the correct context with due regard to spelling, punctuation and grammar. Candidate B clearly has not done enough revision and does not know the straightforward definition of the electron-pair repulsion theory. In what appears to be a rushed answer, the question has been restated. This fails to score. Candidate B's

sketches for methane and ammonia are insufficient to gain the marks. It is particularly important to show the three-dimensional shapes of these molecules. This is best done by using wedge-shaped bonds. Candidate B also loses a mark for not quoting the bond angle in methane but does earn the mark for quality of written communication. Candidate B's score for the four shapes is 5 marks.

Overall, Candidate A scores the full 11 marks, while Candidate B scores 6. Often, chemists do not write in continuous prose but communicate by symbols, equations and specific chemical terms. It is possible to gain the quality of written communication mark without writing at great length. Often as little as two consecutive sentences may be regarded as continuous prose.

Electrical conductivity

Time allocation: 10–11 minutes

Compare and explain the electrical conductivity of magnesium oxide, diamond and graphite. In your answer, you should consider the structure and bonding of each of the materials. (12 marks)

In this question, 1 mark is available for the quality of written communication. (1 mark)

Total: 13 marks

Candidates' answers to Question 11

Candidate A

Electrical conductivity depends on the availability of mobile charge carriers such as free electrons or free ions.

Magnesium oxide is ionic, but the ions are not free to move when solid. However, they are free to move when either molten or aqueous.

Diamond consists of a giant lattice with each carbon atom forming strong covalent bonds to four other carbon atoms. It has no free electrons and is therefore a poor conductor. Graphite also forms a giant lattice, but each carbon atom is only bonded to three other C atoms; hence it has free electrons. Graphite is a good conductor.

Candidate B

	Magnesium oxide	Diamond	Graphite
Conductivity	Poor when solid, good when molten or aqueous Ions only free to move when molten or aqueous	Poor No free electrons	Good Free electrons
Structure	Giant lattice	Giant lattice	Giant lattice
Bonding	Ionic	Covalent	Covalent

e Candidates find open-ended questions like this difficult to answer. It is advisable to plan your answer before you start. Neither candidate did this. The two candidates chose to answer in very different ways. Candidate A decided to write continuous prose, while candidate B tabulated his/her response. Both ways of anwering the question are acceptable. Candidate A's response is excellent, particularly the opening sentence. However, when writing in continuous prose it is easy to miss out simple statements. Candidate A has not stated that magnesium oxide is a poor conductor when solid but a good conductor when molten or in aqueous solution. However, candidate A does imply this and so 1 of the 2 available marks is awarded. It is worth remembering, however, that examiners can only mark what is on the

paper, so even the most obvious statements must be written down. Candidate B's approach is much more systematic and likely to lead to fewer omissions, but in order to gain the quality of written communication mark there must be evidence of continuous prose.

Candidate A scores 10 of the 11 chemistry marks and 1 mark for quality of written communication. Candidate B gained all 11 chemistry marks, but did not gain a mark for quality of written communication.

The mark-scheme for this question is given below. Use it to mark both candidates' answers and see if you agree with the examiner. In questions like this, the examiners often have more marking points than there are marks. When explaining the electrical conductivity of magnesium, diamond and graphite, there is a total of 13 marking points, but a maximum of 11 marks. This means that it is possible to omit 2 marking points and still gain maximum marks.

Magnesium oxide:
- giant; ionic; lattice (2)
- fixed ions in solid; does not conduct when solid (2)
- does conduct when in aqueous solution or molten (1)
- mobile ions in solution or when molten (1)

Diamond or graphite:
- covalent; giant (2)

Diamond:
- no free electrons/charge carriers/all electrons involved in bonding (1)
- does not conduct at all (1)

Graphite:
- layered structure (1)
- delocalised electrons (between layers) (1)
- conducts (by movement of delocalised electrons) (1)

13 marking points, maximum 11 marks

Quality of written communication:
- legible text with accurate spelling, punctuation and grammar; clear, well-organised, using specialist terms correctly (1)